BIOGEOCHEMICAL PROCESSES AT THE LAND-SEA BOUNDARY

FURTHER TITLES IN THIS SERIES

1 J.L. MERO
THE MINERAL RESOURCES OF THE SEA
2 L.M. FOMIN
THE DYNAMIC METHOD IN OCEANOGRAPHY
3 E.J.F. WOOD
MICROBIOLOGY OF OCEANS AND ESTUARIES
4 G. NEUMANN
OCEAN CURRENTS
5 N.G. JERLOV
OPTICAL OCEANOGRAPHY
6 V. VACQUIER
GEOMAGNETISM IN MARINE GEOLOGY
7 W.J. WALLACE
THE DEVELOPMENTS OF THE CHLORINITY/SALINITY CONCEPT IN OCEANOGRAPHY
8 E. LISITZIN
SEA-LEVEL CHANGES
9 R.H. PARKER
THE STUDY OF BENTHIC COMMUNITIES
10 J.C.J. NIHOUL (Editor)
MODELLING OF MARINE SYSTEMS
11 O.I. MAMAYEV
TEMPERATURE-SALINITY ANALYSIS OF WORLD OCEAN WATERS
12 E.J. FERGUSON WOOD and R.E. JOHANNES
TROPICAL MARINE POLLUTION
13 E. STEEMANN NIELSEN
MARINE PHOTOSYNTHESIS
14 N.G. JERLOV
MARINE OPTICS
15 G.P. GLASBY
MARINE MANGANESE DEPOSITS
16 V.M. KAMENKOVICH
FUNDAMENTALS OF OCEAN DYNAMICS
17 R.A. GEYER
SUBMERSIBLES AND THEIR USE IN OCEANOGRAPHY AND OCEAN ENGINEERING
18 J.W. CARUTHERS
FUNDAMENTALS OF MARINE ACOUSTICS
19 J.C.J. NIHOUL (Editor)
BOTTOM TURBULENCE
20 P.H. LEBLOND and L.A. MYSAK
WAVES IN THE OCEAN
21 C.C. VON DER BORCH (Editor)
SYNTHESIS OF DEEP-SEA DRILLING RESULTS IN THE INDIAN OCEAN
22 P. DEHLINGER
MARINE GRAVITY
23 J.C.J. NIHOUL (Editor)
HYDRODYNAMICS OF ESTUARIES AND FJORDS
24 F.T. BANNER, M.B. COLLINS and K.S. MASSIE (Editors)
THE NORTH-WEST EUROPEAN SHELF SEAS: THE SEA BED AND THE SEA IN MOTION
25 J.C.J. NIHOUL (Editor)
MARINE FORECASTING
26 H.G. RAMMING and Z. KOWALIK
NUMERICAL MODELLING MARINE HYDRODYNAMICS
27 R.A. GEYER (Editor)
MARINE ENVIRONMENTAL POLLUTION
28 J.C.J. NIHOUL (Editor)
MARINE TURBULENCE
29 M. WALDICHUK, G.B. KULLENBERG and M.J. ORREN (Editors)
MARINE POLLUTANT TRANSFER PROCESSES
30 A. VOIPIO (Editor)
THE BALTIC SEA
31 E.K. DUURSMA and R. DAWSON (Editors)
MARINE ORGANIC CHEMISTRY
32 J.C.J. NIHOUL (Editor)
ECOHYDRODYNAMICS
33 R. HEKINIAN
PETROLOGY OF THE OCEAN FLOOR
34 J.C.J. NIHOUL (Editor)
HYDRODYNAMICS OF SEMI-ENCLOSED SEAS
35 B. JOHNS (Editor)
PHYSICAL OCEANOGRAPHY OF COASTAL AND SHELF SEAS
36 J.C.J. NIHOUL (Editor)
HYDRODYNAMICS OF THE EQUATORIAL OCEAN
37 W. LANGERAAR
SURVEYING AND CHARTING OF THE SEAS
38 J.C.J. NIHOUL (Editor)
REMOTE SENSING OF SHELF SEA HYDRODYNAMICS
39 T. ICHIYE (Editor)
OCEAN HYDRODYNAMICS OF THE JAPAN AND EAST CHINA SEAS
40 J.C.J. NIHOUL (Editor)
COUPLED OCEAN–ATMOSPHERE MODELS
41 H. KUNZENDORF (Editor)
MARINE MINERAL EXPLORATION
42 J.C.J. NIHOUL (Editor)
MARINE INTERFACES ECOHYDRODYNAMICS

Elsevier Oceanography Series, 43

BIOGEOCHEMICAL PROCESSES AT THE LAND-SEA BOUNDARY

Edited by

PIERRE LASSERRE
Station Marine de Roscoff
CNRS & Université de Paris VI
Roscoff, Nord-Finistère, France

and

JEAN-MARIE MARTIN
Institut de Biogéochimie Marine
CNRS & Ecole Normale Supérieure
rue d'Ulm, Paris, France

ELSEVIER

Amsterdam — Oxford — New York — Tokyo 1986

ELSEVIER SCIENCE PUBLISHERS B.V.
Sara Burgerhartstraat 25
P.O. Box 211, 1000 AE Amsterdam, The Netherlands

Distributors for the United States and Canada:

ELSEVIER SCIENCE PUBLISHING COMPANY INC.
52, Vanderbilt Avenue
New York, NY 10017, U.S.A.

Library of Congress Cataloging-in-Publication Data

Biogeochemical processes at the land-sea boundary.

(Elsevier oceanography series ; 43)
Bibliography: p.
1. Biogeochemical cycles. 2. Seashore ecology.
3. Coastal ecology. I. Lasserre, Pierre. II. Martin,
Jean-Marie. III. Title: The land-sea boundary.
IV. Series.
QH344.B57 1986 574.5'2638 86-16702
ISBN 0-444-42675-2 (U.S.)

ISBN 0-444-42675-2 (Vol. 43)
ISBN 0-444-41623-4 (Series)

Printed in The Netherlands

CONTRIBUTING AUTHORS

M. Bianchi	Laboratoire de Microbiologie Marine, CNRS & Université de Provence, 1331, Marseille, France.
S. Chamroux	Station Biologique de Roscoff, Université de Paris VI & CNRS, 29211 Roscoff, France.
E.T. Degens	Geologisch-Paläontologisches Institut, Bundesstrasse 55, D-2000 Hamburg 13, Federal Republic of Germany.
J.G. Field	Marine Biology Research Institute, Zoology Department, University of Cape Town, 770 Rondebosch, South Africa.
B.R. Folsom	Gray Freshwater Biological Institute, University of Minnesota, Navarre, MN 55392, United States of America.
C.D. Hunt	Graduate School of Oceanography, University of Rhode Island, Kingston, RI 02881, United States of America
V. Ittekkot	Geologisch-Paläontologisches Institut, Bundesstrasse 55, D-2000 Hamburg 13, Federal Republic of Germany.
J. Jednacak-Biscan	Laboratory of Electrochemistry, Rudjer Boskovic Institute, P.O. Box 1013, Zagreb, Yugoslavia.
C.J.M. Kramer	MT-TNO, Department for Marine Research, P.O. Box 57, 1780 AB Den Helder, The Netherlands.
P. Lasserre	Station Biologique de Roscoff, Université de Paris VI & CNRS, 29211 Roscoff, France.
K.H. Mann	Marine Ecology Laboratory, Bedford Institute of Oceanography, Dartmouth, N.C. B2Y 4A2, Canada.
J.-M. Martin	Institut de Biogéochimie Marine, UA 386 CNRS, Ecole Normale Supérieure, 46, rue d'Ulm, 75230 Paris, France.

L.-A. Meyer-Reil	Institut für Meereskunde an der Universität Kiel, Marine Mikrobiologie, 2300 Kiel 1, Federal Republic of Germany.
W. Michaelis	Geologisch-Paläontologisches Institut, Bundesstrasse 55, D-2000 Hamburg 13, Federal Republic of Germany.
C.L. Moloney	Marine Biology Research Institute, Zoology Department, University of Cape Town, 770 Rondebosch, South Africa.
F.M.M. Morel	Department of Civil Engineering, R.M. Parsons Laboratory, Massachusetts Institute of Technology, Cambridge, Mass., United States of America.
J.M. Mouchel	Institut de Biogéochimie Marine, UA 386 CNRS, Ecole Normale Supérieure, 46, rue d'Ulm, 75230 Paris, France.
S.W. Nixon	Graduate School of Oceanography, University of Rhode Island, Kingston, RI 02881, United States of America.
B.L. Nowicki	Graduate School of Oceanography, University of Rhode Island, Kingston, RI 02881, United States of America.
T. Tournié	Station Biologique de Roscoff, Université de Paris VI & CNRS, 29211 Roscoff, France.
D.R. Turner	Marine Biological Association of the U.K., Citadel Hill, Plymouth PL1 2PB, United Kingdom.
P.A. Wickens	Marine Biology Research Institute, Zoology Department, University of Cape Town, 770 Rondebosch, South Africa.
J.M. Wood	Gray Freshwater Biological Institute, University of Minnesota, Navarre, MN 55392, United States of America.

CONTENTS

PREFACE

The connection between terrestrial and oceanic systems is a functional one and the consequences of this linkage on the very large variety of coastal systems are profound. It is widely recognized that many of the major processes which influence the biological properties and chemical forms of elements, and their biogeochemical cycles in the ocean occur at the land-sea boundary, especially in estuaries, coastal lagoons, the coastline and the shelf. Over 80% of living systems and their fisheries take place in near shore waters and the consequential production of organic matter produced triggers off the high level of activity. Hence, the last fifteen years, there has been considerable stimulus to provide framework to evaluate the interactions and effects of human activities.

There is an intuitive belief among scientists that each estuary or lagoon is unique and different. Similarities do exist, however, and this book attempts to highlight some common properties and perhaps some innovative views on biogeochemical processes, and biological fluxes which are of central concern in the understanding of the land-sea boundary.

The contents are based upon lectures given at a Seminar organised as the scientific component of the 17th General Meeting of SCOR, held at the Station Biologique de Roscoff, France, on 22-24 October 1984. The original lectures have been substantially extended and revised in order to give a fuller treatment of the suject. The contributions identify important processes influencing (I) Behaviour of chemical species, (II) Nutrient cycling and mechanisms of organic transformations and (III) Uptake of trace elements by living systems. Special attention was paid to modelling and to the problem of anthropogenic influence on the natural behaviour, with an overview of the methodology and experimental techniques used in laboratory and field research.

The editors wish to acknowledge the sustained support of the Scientific Committee on Oceanic Research (SCOR), the International Association for Biological Oceanography (IABO) and the UNESCO Division of Marine Sciences. Financial support was also provided by the Centre National de la Recherche Scientifique (CNRS 'Pirocean' and 'Piren'), the Université de Paris VI, the Institut français de Recherche pour l'Exploitation de la Mer (IFREMER), and the Ministère des Relations Extérieures. We specifically thank Mrs Bernadette Lasserre for providing excellent copy editing and Mrs Nicole Guyard for secretarial assistance.

The selection of topics is the responsibility of the editors and has undoubtedly been influenced by the many friends and colleagues met through mutual interests in marine biogeochemistry. The editors are most grateful to the authors for their collaboration and to all participants to the SCOR Seminar, in Roscoff, for their involvement in the lively discussions, the results of which are reflected in the final form of contributed chapters. We hope this book will provide insights which will point the way for future research.

Roscoff, April 1986

Pierre Lasserre
Jean-Marie Martin

I

BEHAVIOUR OF CHEMICAL SPECIES

AND

MODELLING

APPROACHES IN CHEMICAL SPECIATION STUDIES IN MARINE WATERS

C.J.M. KRAMER*

Netherlands Institute for Sea Research, P.O.Box 59, den Burg, Texel (The Netherlands)

ABSTRACT

Kramer, C.J.M., 1986. Approaches in chemical speciation studies in marine waters. In: P. Lasserre and J.M. Martin (eds), Biogeochemical Processes at the Land-Sea Boundary. Elsevier, Amsterdam.

After a brief introduction on terminology, this overview summarizes the experimental and theoretical modelling methods applied in trace metal speciation studies, emphasizing the dissolved fraction as defined by 0.45 μm filtration. The experimental approach comprises interactions with organic - and inorganic ligands, speciation schemes, biological experiments and interactions with particles and colloids.

To illustrate one type of speciation research, i.e. the determination of the apparent complexation capacity for copper (CC_{Cu}) and the conditional stability constant (K'), examples are given for three marine areas, viz. the Scheldt estuary, the Southern Bight of the North Sea and the open north Atlantic Ocean. A hypothetical model is presented giving the complexation capacity across the land-sea boundery : from river to ocean.

INTRODUCTION

Interest in trace element speciation studies in natural waters has increased considerably during the last decade. It has become apparent that data on total concentrations of any element rather than on individual well defined chemical entities, are often inadequate to identify transport mechanisms, ultimate fate and toxicity of particular elements to organisms. A study of the different trace metal species and their relative distribution will assist in understanding the chemical processes that take place in the highly reactive estuarine zone and in the open sea. These processes include the rate at which chemical processes take place, the participation in geochemical processes (precipitation/dissolution, adsorption/desorption).

Research in the field of trace metal speciation has been developed in application and improvement of two main directions :

a) experimental techniques to distinguish specific or operationally defined forms of trace elements, either in laboratory studies under controlled conditions or in situ ;

b) mathematical models.

---

*Present adress : MT-TNO, Dept. for Marine Research, P.O. Box 57, 1780 AB Den Helder, The Netherlands.

Recent reviews on chemical speciation are published by e.g. Stumm and Brauner (1975), Florence and Batley (1980) and Leppard (1983) sometimes, with special reference to metal-organic interactions (Mantoura, 1982) or complexation in natural waters (Kramer and Duinker,1984b). Bruland (1983) summarized the distribution and behaviour of trace elements in ocean waters. The occurrence of certain species is largely dependent on the environmental conditions. There exists a strong competition of trace metals with H^+ or major cations like Ca^{2+} and Mg^{2+} in seawater, but also with other trace metals which might form more stable complexes with the ligand in question ; on the other side, many potential ligands or chelators compete for one trace element.

At the land-sea boundary, especially near outlets and in estuaries, a rapid change in environmental conditions occurs during mixing of fresh - and seawater. Reactivity in response to gradients in important parameters like ionic strength, ratios of major components, pH, Eh, turbidity, etc. and the availability of trace elements and complexing agents will affect the distribution of the different species of a trace metal. E.g. in freshwater, cadmium will be present in ionic form, as carbonato- and hydroxide complexes and bound to (organic) humates, in seawater chloro-complexes are dominant (Raspor, 1980b ; Mantoura et al., 1978 b). In the following sections an overview will be presented of the various approaches that have been followed. They are briefly summarized in Table 1. Additionally some experimental results on the complexing behaviour of copper in the marine environment will be presented covering the land-sea boundary : estuary, coastal sea and open ocean, to illustrate the possibilities of this type of research.

TABLE 1

Different approaches in trace metal speciation studies (with some examples).

1- Valency state	(Fe (II)/(III), Cr (III)/(VI)
2 - Metal - inorganic interactions	(Cl^-, CO_3^{2-})
3 - Metal - organic interactions	
a) direct determination of species	(Pb(Me))
b) fuctional groups	($RCOO^-$, R_2NH, RS^4)
c) well defined organic compounds	(amino acids, carbohydrates)
d) experimentally defined groups of organic matter	(humics)
e) model compounds	(EDTA, NTA)
f) natural samples	
4 - Speciation schemes	
5 - Effects of organisms on chemical speciation	
a) bio-assays (uptake, effects)	
b) excretion products	
c) intracellular organic matter	
6 - Interaction with particles and colloids	
a) complexation	
b) ion exchange	
c) adsorption - desorption	
d) precipitation - dissolution	
e) coagulation, flocculation	
7 - kinetics	
8 - chemical modelling	

GENERAL TERMINOLOGY

In the (aquatic) environment elements occur in particulate-, colloidal- and dissolved forms. These forms are usually distinguished by filtration or centrifugation. Traditionally, a 0.45 um (membrane) filter separates the particulate from the dissolved forms. This may result in the passage of colloidal fractions through the filter, classifying colloidal matter incorrectly within the dissolved fraction. Although the interaction between dissolved and particulate (surface) fractions cannot be neglected, it is common in speciation studies to consider the "dissolved" fraction. The dissolved forms of trace elements are mainly present as :

- free hydrated ions;
- ion pairs (where the coordinated water is retained);
- inorganic- and organic complexes (with a covalent bond).

An anion or molecule containing free electron pairs forming a coordination compound (complex) with a central atom (cation, M) is called a ligand (L), containing a donor atom (Stumm and Morgan, 1981). Complex formation with anions or molecules with more than one donor atom is called chelation. Chelators can be important where large organic molecules (e.g. enzymes) have a combination of a favorable tertiary structure and several donor atoms. The complexes thus formed are sometimes specific for a trace element, due to its specific size and charge. Binding of more than one metal to one ligand or chelator results in multi- or polynuclear complexes.

Stepwise and consecutive complex formation reactions between a metal M and a ligand L, can be represented in a general notation by equations (1) and (2) :

$$ML_{n-1} + L \underset{k_d}{\overset{k_f}{\rightleftharpoons}} ML_n \qquad (1)$$

$$M + mL \rightleftharpoons ML_m \qquad (2)$$

The corresponding (thermodynamic) stability constant K and the (thermodynamic) cumulative stability constant (β) are, according to Sillen and Martell (1964, 1971), defined as :

$$K_n = \frac{[ML_n]}{[ML_{n-1}][L]} \qquad (3)$$

$$\beta = \frac{[ML_m]}{[M][L]^m} \qquad (4)$$

They are interrelated by :

$$\beta = K_1 . K_2 . K_3 ... K_n \quad (5)$$

The formation rate is also important for understanding complexation mechanisms in the aquatic environment. In the equation (see also (eq.1)) :

$$-\frac{d[ML_{n-1}]}{dt} = k_f [ML_{n-1}][L] \quad (6)$$

the reaction rate constant k_f for the second order reaction (eq.1) is related to the reactant half life $t_{\frac{1}{2}}$ by :

$$t_{1/2} = \frac{1}{[ML_{n-1}]_i \; k_f} \quad (7)$$

where $(ML_{n-1})_i$ represents the initial concentration of reactant and assuming an approximately 1:1 concentration of the reactants. The notation for the so-called conditional stability constant and conditional rate constant is K' and k_f' respectively.

Analyses in environmental samples, involve reactions with several ligands. The stability- and rate constants thus obtained do not reflect the interaction with a single ligand but rather with the complex multi-ligand system.

DIFFERENT APPROACHES

Determination of different oxidation state

A number of elements can be present in the marine environment in more than one oxidation state, e.g. the metals Fe(II)/(III), Mn(II)/(IV), Co(II)/(III), Cr(III)/(VI), Tl(I)/(III), Sn(II)/(IV) and the metalloids As(III)/(V), Se(IV)/(VI) and Sb(III)/(V). Under aerobic conditions at pH 8 most metals be will thermodynamically stable in the oxidized form. Under anoxic conditions (some bottom waters and sediments) the lower oxidation state may become important.

Kinetic effects or the activity of organisms may result in a deviation from the thermodynamically expected situation. In the euphotic zone the formation of As(III) is related to primary production, may be as a result of detoxifying processes (Andreae, 1979). The oxidation of Mn(II) appears to be kinetically hindered ; while Se(IV) and Se(VI) are, as only element, present in similar concentrations in oceanic waters (Bruland, 1983).

In the case of chromium it is uncertain, whether the oxidation state (III) or (VI) is predominant in seawater. This is mainly due to the observed very slow oxidation rate of Cr(III) (Elderfield, 1970 ; Nakayama et al.1981 ; Gould, 1982 ; van der Weyden and Reith, 1982). Interesting is the possibility of the existence of Cu(I) in marine surface waters. Reduction of Cu(II) would be initiated by photochemically produced H_2O_2. The back oxidation is so slow that Cu(I) species have a sufficiently long lifetime to form complexes, predominantly chlorides; therefore, in fresh- and low salinity waters the Cu (I) content will be low (Zika, 1982 ; Moffet and Zika, 1983).

Considering the micro-environment close to the surface of plankton cells the existence of Fe(II) complexes has been observed, even in an aerobic environment (Morel, this volume). The macro-environment represents evidently not necessarily the same conditions as the micro-environment close to the organisms.

Metal - inorganic interactions

The concentrations of the major inorganic ions in seawater are well known ; in estuarine and coastal areas as well as in interstitial waters anomalies in their constant ratios may occur. The major cations are Na^+, Mg^{2+}, Ca^{2+}, K^+ and Sr^{2+}, the major anions Cl^-, SO_4^{2-}, HCO_3^-, $B(OH)_4^-$, F^- and Br^-. Ion pairs involving these elements and H^+, OH^-, CO_3^{2-}, PO_4^{3-} and SiO_4^{4-}. Under anoxic conditions the S^{2-} - ion and bi- and poly-sulphides become important. A summary of the major ion speciation in seawater is given by Kester et al. (1975).

For species distribution calculations stability constants of metal-ligand complexes are required (e.g. Stiff, 1971 ; Bilinski et al., 1976 ; Wedborg, 1979 ; Simoes Goncalves et al., 1981). The accuracy of many data is insufficient and moreover not all data of the potentially important equilibria are known, resulting in large discrepancies in species distribution suggested by various authors (Tables 2 and 3).

TABLE 2

Calculated copper species distribution in seawater by different authors (in %).

	Reference											
	(1)	(2)	(3)	(4)	(5)	(6)	(7)	(8)	(9)	(10)	(11)	(12)
Species												
Cu^{2+}	1.0	0.7	17	0.6	3	0.7	3	2.3	1	-	5	1.2
$CuOH^+$	1.0	3.7	22	0.3	5	0.8	6	7.7	-	9.2	4	3.1
$Cu(OH)_2$	90.0	-	-	0.3	2	85.0	32	40.7	-	5.3	-	0.4
$CuCO_3$	7.7	21.6	49	3.5	87	5.6	50	46.1	11	71.4	79	56
$Cu(CO_3)_2^{2-}$	0.1	-	-	-	-	0.1	6	-	-	10.1	8	-
CuHCO	0.2	-	-	3.5	-	0.2	1	0.3	-	-	-	1.1
$CuCl^{+3}$	0.1	5.8	10	95.5	1	0.1	1	2.5	9	0.6	-	0.3
$CuCl_2$	-	1.6	-		<1	-	-		3	-	-	-
CuOHCl	-	65.2	-	-	1	-	-	-	75	-	-	0.2
$CuSO_4$	0.2	-	2	0.1	1	0.1	<1	0.4	-	-	-	-
CuL	-	0.08	-	-	-	10	-	-	-	-	-	40

1, Zirino and Yamamoto (1972) ; 2, Dyrssen and Wedborg (1974) ; 3, Ahrland (1975) ; 4, Sibley and Morgan (1975) ; 5, Paulsen (1978) ; 6, Mantoura et al. (1978) ; 7, Dyrssen and Wedborg (1979) ; 8, Sylva and Florence (1980) ; 9, Whitfield and Turner (1980) ; 10, van den Berg (1983 b) ; 11, Morel and Morel-Laurens (1983) ; 12, Zuehlke and Kester (1983).

There seems to be a general agreement on the major cadmium, zinc, lead and copper species in seawater. Species with an abundancy >10% are considered $CdCl^+$, $ZnCl^+$ and Zn^{2+}, $PbCO_3$ and $CuCO_3$ (Table 3).

TABLE 3

Predominant species (> 10 %) of cadmium, zinc, lead and copper in seawater.

	Species								
Cadmium	$CdCl^+$	$CdCl_2$	$CdCl_3^-$						
	+	–	–						1
	+	+	–						2
	+	+	+						3
	+	+	+						4
	+	+	+						5
	+	+	+						6
Zinc	Zn^{2+}	$ZnCl^+$	$ZnCl_2$	$ZnOHCl$	$ZnCO_3$	$Zn(OH)_2$			
	+	–	–	–	–	–			1
	+	–	–	–	–	+			2
	+	+	+	+	–	–			3
	+	+	–	–	–	–			4
	+	+	–	+	–	–			5
	+	–	–	–	+	–			6
Lead	$PbCl^+$	$PbCl_2$	$PbCO_3Cl$	$PbCO_3$	$Pb(CO_3)_2^{2-}$	$PbOH^+$			
	–	–	–	+	+	–			8
	+	–	–	+	–	–			2
	+	+	–	–	–	+			3
	–	–	–	+	+	–			9
	–	–	–	+	+	–			10
	–	+	–	+	–	+			16
Copper	Cu^{2+}	$CuCl^+$	$CuOHCL$	$CuCO_3$	$Cu(CO_3)_2^{2-}$	$Cu(OH)^+$	$Cu(OH)_2$	CuL	
	–	–	–	–	–	–	+	–	2
	–	–	+	+	–	–	–	–	3
	+	+	–	+	–	+	–	–	4
	–	+	–	–	–	–	–	–	11
	–	–	–	+	–	–	–	–	12
	–	–	–	–	–	–	+	+	13
	–	–	–	+	–	–	+	–	5
	–	–	–	+	–	–	+	–	14
	–	–	+	+	–	–	–	–	9
	–	–	–	+	+	–	–	–	15
	–	–	–	+	–	–	–	–	6
	–	–	–	+	–	–	–	+	17

1, Baric & Branica (1967) ; 2, Zirino & Yamamoto (1972) ; 3, Dyrssen & Wedborg (1974) ; 4, Ahrland (1975) ; 5, Dyrssen & Wedborg (1980) ; 6, Morel & Morel-Laurens (1983) ; 8, Stumm & Bilinski (1972) ; 9, Whitfield & Turner (1980) ; 10, Sipos et al. (1980) ; 11, Sibley & Morgan (1975) ; 12, Paulsen (1978) ; 13, Mantoura et al. (1978a) ; 14, Sylva & Florence (1980) ; 15, van den Berg (1983b) ; 16, Nürnberg (1983) ; 17, Zuehlke & Kester (1983).

Few trace element species can be analysed in natural samples directly. Ion selective electrodes (ISE) allow measurements of metal ion activity (Cu^{2+}, Cd^{2+}) ; however, their use in the marine environment is limited due to low sensitivity and interference by Cl^-.

Metal-organic interactions

The mixture of organic constituents in the marine environment is extremely complex. Their origin is partly terrigenuous ; many compounds are produced in the marine environment itself. The different sources and their relative importance for the complexation of trace metals in estuaries, coastal seas and open ocean are : riverine input, runoff from the coastal zone, resuspension, import of water mases, atmospheric input and in situ biological production.

Although many groups of compounds can be analysed in seawater qualitatively and quantitatively, e.g. fatty acids, sugars, amino acids, sterols, vitamins (Degens and Ittekot, 1983 ; Laane, 1982), their sum covers only roughly 10-20% of the total DOC. Most compounds are thus unidentified. They include excretion- and decay products of organisms, and decomposition products of organic matter in various stages of decay, subject to transformations and recombinations. The composition and distribution of the dissolved organic matter varies therefore widely.

Many of these organic compounds contain one or more functional groups ; these can form complexes with trace metals. This may affect the speciation of these trace elements significantly.

Predictions based on stability constants for metal ions with some types of functional organic groups suggest that, in seawater, only few trace metals form complexes with these organic compounds to a considerable extent. These elements are copper, iron and, to a lesser extent, zinc, lead and probably nickel. Copper, being the most likely to form organic complexes has been investigated extensively. There is no agreement on which fraction of total copper is organically-bound, it ranges between 0 and nearly 100% (Mantoura, 1982 ; van den Berg, 1984) ; Mills and Quinn (1984) found a decrease (70-14%) with increasing salinity for the Naragansett Bay estuary.

It is interesting to see that few model studies (Table 2) have considered copper-organic interactions (Dyrssen and Wedborg, 1974 ; Mantoura et al., 1978 a; Zuehlke and Kester, 1983), although their appreciable concentrations in seawater seems to be well documented. Various attempts have been made to improve the knowledge on trace metal - organic interactions.

Direct determination of species. With a combination of techniques (e.g. GC and AAS) several distinct organic species càn be isolated and determined, e.g. organo-lead, -mercury and -tin compounds (Chau and Wong, 1983). Also the distinction of several organic and inorganic species of As (Braman et al., 1977 ; Andreae, 1977) and Sb (Andreae, 1983) have been reported.

Functional groups. The study of well defined ligand sites ($RCOO^-$, R_2NH, RO^-, RS^-, $-NH_2$) is based on the preference of the respective trace metals for the different donor atoms, as summarized by Stumm and Brauner (1975). Studies of relative simple organics give insight into the much more complex organics like humic substances and proteins (Gamble and Schnitzer, 1973).

Well-defined organic compounds. Some of the relatively few organic molecules in seawater with a known structure have been examined for their binding ability with trace metals, e.g. amino acids (Simoes Goncalves et al., 1982, 1983 ; Valenta et al., 1984) and phtalic-, citric- and salycilic acid (Stumm and Brauner, 1975).

Experimentally defined groups of organic matter. These types of organic matter can be divided into bio- and geo-polymers. Bio-polymers (produced by living organisms) are e.g. polysaccharides, polypeptides and proteins, enzymes, pigments and vitamins (Reuter and Perdue, 1977). Geo-polymers originate from decay and decomposition processes. They can be divided into operationally defined subgroups :

- humic acid, only soluble in alkaline solutions;
- fulvic acid, soluble in acidic and alkaline solutions;
- humin, insoluble in acidic and alkaline solutions.

These groups of organics with molecular weights ranging 500 to > 100,000, often contain many complexing sites, predominantly carboxylic-, phenolic- and amine type functional groups. Another experimentally defined group of organic compounds able to complex trace metals, is yellow substance (Gelbstoffe), organic matter showing fluorescent behaviour when excitated with blue light (Shapiro, 1964 ; Ghassemi and Christman, 1968).

A problem for both humic- and yellow substances is that for these groups of experimentally defined components of different sources, each analysis will be ambiguous in terms of relative composition and molecular weight distribution. Additionally it appears that almost every scientist working in this field has developed his own extraction procedure (Weber and Wilson, 1975 ; Mantoura and Riley, 1975 a; Schnitzer, 1976 ; Stuermer and Harvey, 1977). Different extraction times and -procedures result in different compositions of the organic constituents (Laane and Kramer, 1984). Soil humic- and fulvic acids, often used for studies on the interaction with trace elements, and those derived from water have certainly not the same composition and contain not the same distribution of functional groups. Therefore, results should be compared with care (Buffle, 1980 ; Buffle et al., 1984).

Almost all trace metals of interest have been investigated with respect to their complexation with humic- and fulvic acids and to their stability constants (e.g. Ernst et al., 1975. Mantoura et al., 1978 b; Bresnahan et al., 1978 ; Takamatsu and Yoshida, 1978 ; Raspor et al., 1984). Results were summarized and compared by Mantoura (1982) and Buffle et al. (1984).

Model compounds. Extensive studies have been carried out concerning the binding of metals (Zn, Cd, Pb and Cu) with model ligands like EDTA, NTA or DTPA (e.g. Raspor and Branica, 1975 ; Raspor et al., 1977, 1978 ; Stolzberg, 1977 ; Plavsic et al., 1980), and with polyelectrolytes (PAA, PMA) as model compounds for humic substances (van Leeuwen et al., 1981 ; Cleven, 1984).

Natural samples. The species distribution of a trace element in the natural environment is affected by the total assemblage of all inorganic and organic ligands. An approach has been developed where the net binding effect of all these ligands is studied by the determination of the apparent complexation capacity e.g. for copper (CC_{Cu}). This is a measure for the concentration of organic ligands (L) able to bind ionic metal into non-labile complexes. The apparent complexation capacity can be determined by titration with an ionic trace element. Conditional stability constants (K') can be calculated from the titration curve. However it should be realized that more than one ligand is involved. Several analytical procedures have been developed for the determination of the CC_{Cu} and K' (see later).

Speciation schemes

Several authors have designed a speciation scheme, i.e. a sequence of analytical techniques to distinguish the various chemical forms in solution (0.45 or 0.2 μm filtration). The sample is divided into several fractions, according to size/density (ultra-filtration, (ultra)-centrifugation, dialysis), charge (electrophoresis), electrochemical behaviour (distinction between labile and non labile forms (ASV)) or chromatographic characteristics (Chelex resin, XAD-2, Sephadex). The distinction may include UV-oxidation. The schemes result in 2-9 fractions. Examples of different speciation schemes are given in Table 4. The significance of the individual fractions is not easy to understand as they are usually experimentally defined. Each scheme results in its own separation products. This should be taken into account when compairing results for the natural environment obtained with different speciation schemes.

TABLE 4

Use of different separation techniques applied to trace metal speciation schemes (n = number of fractions).

REFERENCES	TECHNIQUES										n
	1	2	3	4	5	6	7	8	9	10	
Benes & Steinnes (1974)			*	*		*		*			5
Guy & Chakrabarti (1976)		*	*								8
Smith (1976)	*	*			*						5
Batley & Florence (1976)	*				*		*	*			7
Hart & Davies (1977)	*		*					*			3
Oda & Okabe (1977)	*	*									5
Sugai & Healy (1978)	*				*					*	4
Sugimura et al.(1978)	*							*	*		4
Harrison & Laxen (1980)		*			*			*			9
Hasle & Abdullah (1981)	*	*			*						8
Nürnberg (1984)	*				*		*				3

1, 0.45 or 0.2 μm filtration ; 2, ultra-filtration ; 3, dialysis ; 4, (ultra)-centrifugation ; 5, electrochemical distinction labile - non labile (ASV) ; 6, electrophoresis ; 7, UV-oxidation ; 8, chelex resin ; 9, XAD-2 ; 10, sephadex.

Biological interactions

Many studies have been concerned with uptake by and effects of trace elements on organisms, sometimes in relation to different chemical species of the element concerned (e.g. Langston and Bryan, 1984). As an example, the extent to which copper is complexes has been found to control its toxicity to organisms (Sunda and Guillard, 1976 ; Anderson and Morel, 1978 ; Jackson and Morgan, 1978).

Of concern for this paper is the effect of organisms on trace metal speciation. Most studies have been carried out in batch cultures, flow-through systems or large plastic bags, usually containing one or a few diatom-, algal- or phytoplankton species. In situ data are scarce. The impact of living marine organisms on the complexation of trace metals in solution, can be explained in different terms : adsorption or complexation to the cell wall and complexation with excretion products and/or intracellular fluids. Decay products have been discussed before.

The adsorption of copper onto phytoplankton and macrophytes was reported by Hunt and Fitzgerald (1983). Excretion products of organisms may be rich in complexing agents, and the production of exudates might partly be induced by high trace metal concentrations. Johnston (1964) and Khailov (1964) already mentioned their possible chelating effects. Which were experimentally confirmed in cultures for phytoplankton and algae (Gnassia-Barelli et al., 1978 ; Swallow et al., 1978 ; Ragan et al., 1979 ; Mc Knight and Morel, 1979 ; Sueur et al., 1982 ; Collier and Edmond, 1984 ; Seritti et al., 1985). Diatoms were also found to produce chelating organics (Fischer and Fabris, 1982 ; Imber and Robinson, 1983 ; Imber et al., 1984). Surface active compounds can be chelators as well (Wallace, 1982) and have a strong relation with the activity of plankton blooms, both in batch culture and in the field (Zutic et al., 1981).

Interactions with particles and colloids

Reactions of dissolved species with particulate and colloidal suspended matter include adsorption/desorption, complexation, ion-exchange, precipitation/dissolution, coprecipitation during coagulation and flocculation (Morgan, 1966 ; Stumm and Morgan, 1981 ; Parks, 1975). These processes are particularly important at the land-sea boundary in estuaries (Duinker, 1980 ; Martin et al., this volume). The interaction with particles > 0.45 µm is not discussed here.

Colloids. Colloids include particles with hydrophobic, hydrophilic and intermediate forms with a size range 1 - 400 nm. Both organic (including macromolecules) and inorganic (hydrolyzed silica and metal oxides) colloids occur in the marine environment (Sigleo and Helz, 1981). Their surfaces often contain suitable sites for interactions with trace metals (adsorption, complexation). In the marine environment all particles have a negative surface charge (Neihof and Loeb, 1972 ; Hunter and Liss, 1982). Increase of the electrolyte concentration decreases the stability of the colloidal particles. As a result the instable colloids will agglomerate to larger particles and eventually to large flocs (Stumm and Morgan, 1981 ; Mayer, 1982). As these processes already take place at slightly increased electrolyte concentration, the lower salinity range of an estuary is the most important area for coagulation processes including organics (Sholkovitz, 1976, 1978). The coagulates and flocs formed may act as scavengers for trace elements. Aggregation of fulvic acid seems to be enhanced by copper ions, as was found by Underdown et al. (1981) and Gamble et al. (1984) using laser light Rayleigh scattering.

Relatively little is known about natural colloids. Their particle size distribution and the changes therein under variable natural conditions (estuaries) is poorly documented. One of the problems is their very small size and their susceptibility to changes during sample handling : filtration, storage, etc. (Sigg et al., 1984). Concentration techniques (centrifugation and/or ultrafiltration) are applied for collection of colloids to study their chemical composition, size distribution and their chemical reactions with metals (Lammers, 1967 ; Boonlayangoor et al., 1979 ; Kramer and Duinker, 1984a).

The fulvic- and humic acids in rivers and estuaries are an important part of the colloidal fraction, as was demonstrated with small angle X-ray scattering (Thurman et al., 1982), it is uncertain whether high molecular weight organics (e.g. humics) are in true solution or in colloidal forms (Stumm and Morgan, 1981). Iron has also been found to be predominantly present in colloidal particles in estuaries, e.g. in the Beaulieu estuary (Moore et al., 1979), the Patuxent estuary (Sigleo et al., 1982), the Yarra estuary (Hart and Davies, 1981) and several rivers in SW England (Mill, 1980).

Important complexation and/or sorption reactions between colloids and trace elements may occur. In studies on the interactions of copper with EDTA and γAl_2O_3 as model-ligand and -particles, the apparent complexation capacity was increased by the addition of the particulates (Plavsic et al., 1980). This was also observed for the influence of MnO_2-colloids on the complexation capacity for lead (Sigg et al., 1984).

In natural systems therefore part of the complexation capacity might be caused by colloidal material. This was demonstrated in experiments on the complexation capacity of samples from the Scheldt estuary at different salinities, determined as function of several concentration steps, using a hollow fiber ultrafiltration set up with a theoretical cut off of MW 5000 (Kramer and Duinker, 1984a).

The CC_{Cu} appeared to be linearly dependent on the concentration factor only in the upper part of the estuary. No increase in the CC_{Cu} could be observed after concentration at higher salinities. The riverine colloidal material apparently coagulates to particles and flocs at increasing salinity in the upper part of the estuary. These flocs are retained by the 0.45 µm filter, thus no longer contributing to the complexation capacity of the "dissolved" fraction. This explains the non conservative behaviour of the CC_{Cu} in this part of the estuary (see later). Samples taken from the north Atlantic Ocean did not show an increase in CC_{Cu}, even after a concentration factor 200 times.

Reaction kinetics

Recently, Pankov and Morgan (1981a,b) emphasized the importance of various mechanisms for regulating kinetics in the aquatic environment. Examples showed the wide range of first- and second order rate constants (k_f) and half lifes ($t_{\frac{1}{2}}$) for different reactions that might take place in natural waters. The rate constants for several first order trace metal hydrolysis reactions, second order redox- and complexation reactions of interest for aquatic studies are summarized by Hoffmann (1981). His comparison of kinetic data on the oxidation of HS^- under only slightly different conditions shows considerable variations ; e.g., $t_{\frac{1}{2}}$ ranges from 7 -600 min for seawater media.

The rate of a given reaction in the marine environment depends on temperature and pH, concentration of the reactants and competing compounds and effects of (auto)catalysis. This may have severe effects in regions where large gradients in these parameters occur (estuaries, coastal regions), e.g. the change of $t_{\frac{1}{2}}$ for the oxidation reaction of Fe^{2+} with pH; found was 10^{-1} s at pH 6 and 10^3 s at pH 8 (Stumm and Lee, 1961 ; Roekens and van Grieken, 1983). Catalysis will influence (usually improve) the speed of the overal reaction. Also surfaces are known as catalysts. Evidence has been found that autocatalysis of Fe(II) can take place under certain circomstances by Fe(III) precipitates (Sung and Morgan, 1980).

The kinetics of the oxidation of Cr(III) and Cu(I) have been discussed before. Cr(VI) is reduced by dissolved organic matter, the slow re-oxidation resulting in a large enough $t_{\frac{1}{2}}$ for an existence of Cr(III). Also the existence of Cu(I) in seawater is a steady state between the reduction- and back-oxidation reactions. The lifetime is dependent on pH, PO_2, complexing ligands and redox intermediates such as H_2O_2 (Moffet and Zika, 1983).

Taking advantage of their high sensitivity and selectivity for easily reducible trace metal species, several electrochemical techniques are used to obtain kinetic data of complexation reactions. Application of dc-ASV at a slowly dropping mercury electrode (SDME) the effect of different model solutions on the time dependence of the chelation of Cd(II) with EDTA was studied (Rasport et al., 1977, 1980a). The reaction rate constant k_f ranged from unmeasurably fast in 0.59 M NaCl to 4.2×10^2 $M^{-1}. s^{-1}$ in 0.01 M $CaCl_2$. The value in natural seawater was $(3.6 \pm 0.5) \times 10^2$ $M^{-1}. s^{-1}$, the $t_{\frac{1}{2}}$ for the second order reaction was found to depend on the initial concentration of the reactants, in accordance with theory (see eq. 7). Depending on the parameters used in ASV analysis the values of k_f which can be determined are 10^2 - 10^6 $M^{-1}. s^{-1}$, if one asumes an initial concentration range of 10^{-6} -10^{-8} M (Valenta, 1983).

The application of a rotating ring-disc electrode in ASV resulted in a k_d for the Cd-EDTA complex of 12.4 s^{-1} ; the k_f determination was less reliable however (Shuman and Michael, 1975). With a rotating disc electrode applied at different rotation speeds, the same authors (Shuman and Michael, 1978) determined a first order dissociation rate constant for copper chelates in coastal Atlantic waters (in the order of 2 s^{-1}). Based on a kinetic criterion the chelation of copper in these samples was estimated.

Titration curves of natural samples for complexation capacity determinations can be influenced by the kinetics of the complexation reaction(s). If the time allowed for equilibration after spiking of the sample is short with respect to the time required for establishment of chemical equilibrium, a deviation from the two intersecting lines will occur (Duinker and Kramer, 1977). In a Ruzic plot (Ruzic, 1982) M_F vs. $M_F/(M_T-M_F)$, where M_T and M_F are total and free metal respectively, a typical change of slope bulb appears under these circomstances, giving information on the rate constant (Ruzic and Nicolic,1982). This method was applied to North Atlantic surface water samples, for the complexation of ionic copper by natural ligands. These results indicate slightly different conditional rate constants for different areas, k_f' ranging from 3.6×10^4 in the north to $1.4 \times 10^4 \cdot M^{-1} s^{-1}$ in more southern areas, Table 7 (Kramer, 1986).

In many experiments chemical equilibrium is assumed, however dynamic non-equilibrium processes in seawater may result in products different from those expected under equilibrium conditions. Equilibria can be complex and may involve several types of reactions simultaneously. Kinetics can therefore affect speciation of trace elements and should be taken into account.

Chemical modelling

In chemical modelling, physicochemical principles are applied for the interpretation of natural hydrogeochemical systems (Jenne, 1979). In most cases the presence of chemical equilibrium of the system is assumed. This is, of course, not always the case in natural systems but this approach is much less complicated than an approach for a dynamic system. Certain dynamic situations can be treated however by chemical equilibrium theory, especially in cases where very slow or very fast reactions occur (Pankov and Morgan, 1981b). The equilibrium models attempt, usually by iterative calculations, to find the most stable situation for a given set of pressure, temperature and composition of a system or, in thermodynamic terms, to minimize the value for the Gibb's free energy.

Calculations can be carried out by a) minimizing the free energy function (free energy minimizing method) or b) solving a set of non-linear equations consisting of equilibrium constants and mass balance constraints (equilibration constant method). The latter method is preferred, as the values of the equilibration constant data are more reliable than the free energy data (Nordstrom and Ball, 1984). The models try to predict a species distribution of a system given a set of components together with temperature and pressure. The result will be a first approximation of the real system, which might yield useful information for the natural situation. The validity of the results, or how close is the real (natural) system approximated by the model predictions is, apart from the quality of the theoretical model (interactions involved, refinement of calculation), largely dependent on the accuracy and reliability of the data used, and the choice of physicial and chemical components involved. The causes of discrepancies between model and real systems are summarized by Stumm and Morgan (1981).

Some results of application of theoretical models to trace metal speciation were presented in the section on metal-inorganic interactions (tables 2 and 3) ; a collection of papers dealing with chemical modelling in aqueous systems, including speciation, sorption, solubility and kinetics was edited by Jenne (1979). Recently Nordstrom and Ball (1984) summarized 58 aqueous chemical equilibrium computer programs of which 19 were dealing with trace metals.

STUDIES ON NATURAL SAMPLES

Even before an experiment or analysis is carried out, many factors may affect the speciation of trace elements in the sample under investigation.

Sampling, storage and experimental conditions

Sampling. At the trace levels of most constituents in seawater it is difficult not to contaminate the sample during the sampling procedure, not only for the trace element(s) in question, but also for possible ligands. Water samplers and hydrographic wires have been intercompared (Batley and Gardner, 1977 ; Bruland et al., 1979 ; Bewers and Windom, 1982; Spencer et al., 1982 ; Bewers and Windom, 1983), surface sampling has to be carried out away from the ship, upstream of a slowly moving small raft (Mart, 1979b).

Filtration. To separate the dissolved from the particulate fraction filtration or centrifugation will be necessary. This causes a severe risk of contamination ; it is therefore often not carried out with open ocean samples, where the concentrations of suspended material are low. The presence of phytoplankton or a variable concentration of suspended matter affects the total concentration and a comparison of samples can thus become difficult. In speciation studies the presence of particles may influence the results even more (complexation, adsorption), therefore filtration over acid washed membrane or screen filters in an appropriate filtration apparatus is recommended for all natural samples (Bewers et al., 1985). High pressure during filtration should be avoided : ruptured (plankton-) cells will contribute organic matter, nutrients and trace metals to the solution. A pressure $<$ 25 kPa is recommended (Florence and Batley, 1980).

Clogging of filters when high loads of suspended matter are present (rivers, estuaries, coastal waters), will trap much smaller particles (including colloids) than corresponds with the initial filter pore size, causing a severe influence on the speciation of trace elements (Kennedy et al., 1974 ; Danielson, 1982). Filtration of a limited (known) amount of water or to terminate filtration once a required over-pressure has built up above the filter (Laxen and Chandler, 1982) might improve the filtration procedure. Use of cross-flow or tangential-flow filtration instead of the normally used dead-end filtration might result in a better defined dissolved fraction, with the additional advantage of increased filtration speed. However as far as we known, no (trace metal) contamination free system for limited volumes ($\pm$ 1 liter) is available yet.

Centrifugation. Particle separation by centrifugation is based on particle density and consequently, centrifugated samples differ from the experimental definition of the dissolved fraction, as based on filtration. Centrifugation may result in a relative enrichment of less dense particles in the dissolved fraction, especially when a flow-through centrifuge is used. There is no garantee that all particles are removed from the dissolved fraction. Small but significant differences were found between filtration and centrifugation techniques (Duinker et al., 1979). Centrifugation has been applied to separate samples into different fractions (Benes and Steinnes, 1974 ; Salbu, 1981). In the field of biochemistry (ultra-)centrifugation has become a powerful tool for separations, which might be applied in trace metal speciation studies.

Storage. Samples for total trace metal content are frequently stored after acidification to pH 1. This method is of course not possible in speciation analyses, where direct analysis is the optimal choice. Unfortunatly, this is not always possible. Small differences due to prolonged storage have been reported (Batley and Gardner, 1977). For storage, best results can be expected using polythene or Teflon bottles, acid cleaned and equilibrated with (sea)water prior to use, to eliminate adsorption to the wall (Mart, 1979a). Samples should be stored in the dark at about 4°C ; deep freezing will result in changing of equilibria (Batley and Gardner, 1977 ; Krom and Sholkovitz, 1978).

Experimental conditions. Conditions during analysis should reflect natural system conditions as close as possible, or they should be controlled in such a way that the results can be interpreted properly. Due to the relatively large H^+ concentration - with respect to trace elements - a change in pH can have serious influence on trace element species distribution (Zirino and Yamamoto, 1972). Two major methods are used for pH control. The first involves the application of a pH-stat, using CO_2 as convenient reagent (Kramer and Manshanden, 1982). In the second method, a buffer is added to the sample. To avoid serious changes in the distribution of the species (most buffers contain complexing constituents) the use of special buffers is recommended. Jardim and Allen (1984) used PIPES (Piperazine-N-N'-bis(2-ethanesulfonic acid)) at pH 6.6, which has a low stability constant with copper. Because (almost) no chemicals are added and the measurement is carried out at natural pH, the first method seems to be preferable.

Complicated biological systems (bioassays) at trace element concentration levels typical for offshore waters, are subject to serious danger of contamination. Without extreme precautions e.g. Carpenter and Lively (1980) and Fitzwater et al. (1982) found the toxic effect (inhibition of primary production) of contamination by the incubation bottles. Effects of adsorption to walls and particulate matter (sediment) should not be underestimated. Use of clean lab techniques and regular check of the trace element concentrations throughout the (biological) experiments is necessary to get an indication of the actual concentration and possible distribution of the different elements. Depending on the type of experiments it could be possible that other parametes should be known or even controlled : pE, PO_2, ionic strength, temperature, DOC etc.

Complexation capacity and conditional stability constants in marine waters.

Techniques. Methods which are useful for identification of individual species were classified by Stumm and Brauner (1975). Several techniques have been developed for studies of the apparent complexation capacity (CC_{Cu}). The technique most often applied is differential pulse anodic stripping voltammetry (DPASV) at various electrode types (Duinker and Kramer, 1977 ; Srna et al., 1980 ; Plavsic et al., 1982 ; Seritti et al., 1983 ; Kramer and Duinker, 1984a ; Kramer, 1985, 1986). Ligand competition with catechol followed by cathodic stripping voltammetry (CSV) detection of Cu-catechol complex ions makes use of another electrochemical technique (van den Berg, 1984). Other techniques include the solubilisation of copper sulphide (Kerr and Quinn, 1980), MnO_2-adsorption (van den Berg and Kramer, 1979), ligand exchange with EDTA (Hirose et al., 1982), ion exchange (Stolzberg and Rosin, 1977 ; Crosser and Allen, 1977 ; Buffle et al., 1977 ; Mc Crady and Chapman, 1979 ; Buffle et al., 1980 ; Wood et al., 1983) and the use of biological response (bioassay) studies (Gillespie and Vaccaro, 1978 ; Brand et al., 1983 ; Sunda and Ferguson, 1983). Ion selective electrodes (ISE) have not been used in seawater, due to their lack in sensitivity. Other techniques like dialysis (Truitt and Weber, 1981 ; del Castilho et al., 1984), gel filtration (Mantoura and Riley, 1975 b ; Evans et al., 1979) and fluorescence quenching (Ryan and Weber, 1982 ; Berger and Ewald, 1984) have been applied to fresh waters only. The techniques for the determination of the CC_{Cu} are described and compared by Hart (1981) and Neubecker and Allen (1983). A summary of the CC_{Cu} and K' in marine waters is given by Kramer (1986).

Other techniques applied to speciation studies are potentiometric stripping analysis (PSA) (Jagner and Aren, 1982), pseudopolarography (Brown and Kowalski, 1979), electrophoresis (Röhl,1982), several chromatographic techniques, XAD-1 and -2 (Mackey, 1982) and SEP-PAK-C_{18} (Mills et al., 1982 ; Plavsic and Branica, 1985),HPLC (Mackey, 1983), size exclusion chromatography (Gardner et al., 1982), equilibration diafiltration (Lee, 1983) and light scattering (Underdown et al., 1981).

CC_{Cu} and K' determination by DPASV analysis. To illustrate the possibilities of speciation studies, an example of the application to natural waters will be presented in this section, with special reference to the determination of the complexation capacity for copper of marine waters.

Voltammetric measurements were made with a PAR 174 polarographic analyser with a PAR 315 electroanalysis controller (EG & G) and a jet-stream mercure film electrode (JMFE, Magjer and Branica, 1977). The mercury film was applied to the Glassy Carbon surface (t_{dep} = 10 min, 8 x 10^{-5} M Hg, 50 Hz) and formed prior to the analyses. An Ag/AgCl (sat. KCl) reference electrode (Metrohm) and a platinum wire counter electrode were fitted in a 100 ml FEP Teflon cell. Typical setting for CC_{Cu} titrations was t_{cond} = 90 s, t_{dep} = 180 s, E_{cond} = 0 mV, E_{dep} = -900 mV, scan speed 10 mVs^{-1} and a modulation amplitude of 50 mV. All CC_{Cu} and K' results are based on the direct titration method with copper (Duinker and Kramer, 1977) and calculated according to the method proposed by Ruzic (1982). For the analytical procedure and further details see Duinker and Kramer (1977) and Kramer (1986).

Scheldt estuary. Surface water samples in the Scheldt estuary were collected at high water slack at 9 stations (Fig. 1). Salinities covered were in the range 0.7 - 30.3 x 10^{-3}. The CC_{Cu} data, based on triplicate measurements are illustrated in Fig. 2. The plot does not reflect conservative behaviour, unlike e.g., dissolved fluorescing organic matter. A rapid decrease is observed between salinities 0.7 and 10 x 10^{-3}. It could be concluded from ultra-filtration eperiments (see before), that colloids and/or large organic molecules play an important role in the complexation of ionic copper. These are removed in the early stages of estuarine mixing.

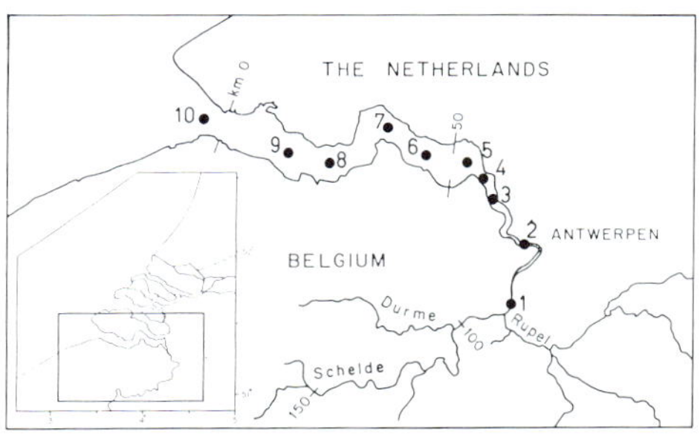

Fig. 1. Sampling stations in the river Scheldt estuary.

TABLE 5

Complexation Capacity (CC), conditional stability constant (K'), temperature, pH, oxygen and suspended matter content in the river Scheldt estuary at different salinities.

Station	S(x10^3)	CC(nMCu^{2+})	log K'	T(°C)	O_2(ppm)	pH	SM(mgl^{-1})	mFl
1	0.7	299	7.4	7.2	3.2	7.40	83	93
2	3.0	245	7.7	7.3	8.1	7.62	65	102
3	5.9	221	7.5	6.8	9.7	7.76	117	92
4	9.2	173	7.3	6.4	11.8	7.85	67	78
6	14.9	159	7.4	5.1	11.3	7.95	45	67
7	18.7	146	7.4	5.4	11.0	8.03	54	58
8	21.7	152	7.3	5.4	10.4	7.99	47	50
9	25.2	130	7.2	5.5	10.6	8.00	46	35
10	30.3	108	7.4	5.6	10.6	8.00	47	25

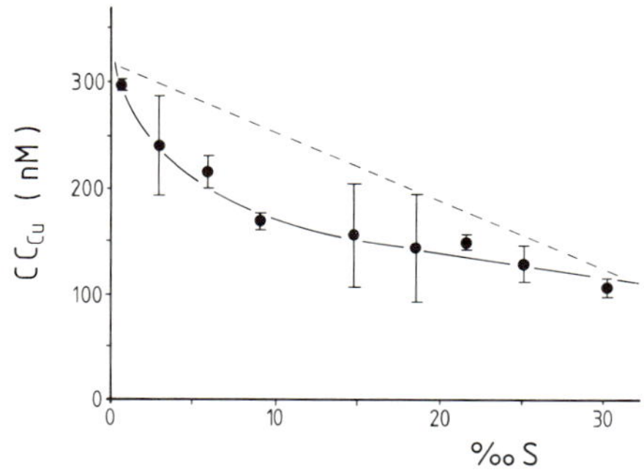

Fig. 2. Means of triplicate analysis of apparent Scheldt estuary copper complexation capacity in the river Scheldt estuary at various salinities.

The conditional stability constants calculated for the different salinity ranges are given in Table 5. It looks as if in the more saline samples complexes with lower K' are formed.

Southern North Sea. Surface watersamples were collected in the southern part of the North Sea, in coastal waters and offshore in October 1982 (Fig. 3). In Table 6 the CC_{Cu} and log K' are presented. It appears that coastal waters (stations 1 and 2) show a higher complexation capacity (82 nM Cu^{2+}) then the samples with less contribution of river water (stations 3 and 4) with a CC_{Cu} of 44 nM Cu^{2+}, with an intermediate value for the station in the German Bight.

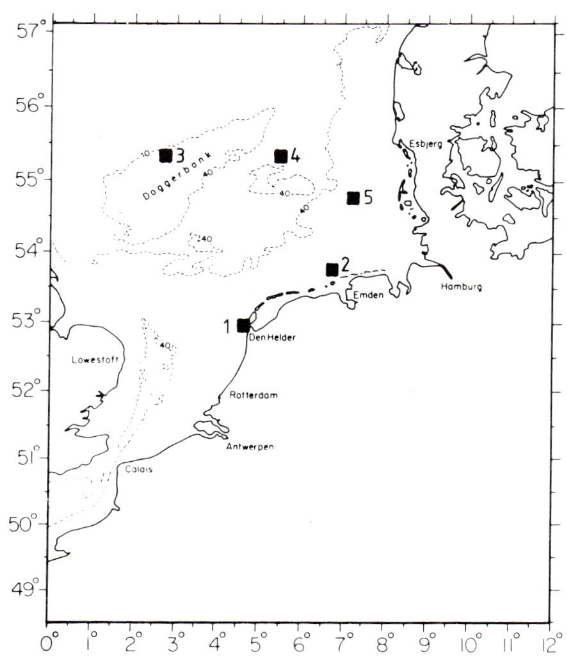

Fig. 3. Sampling stations in the Southern North Sea.

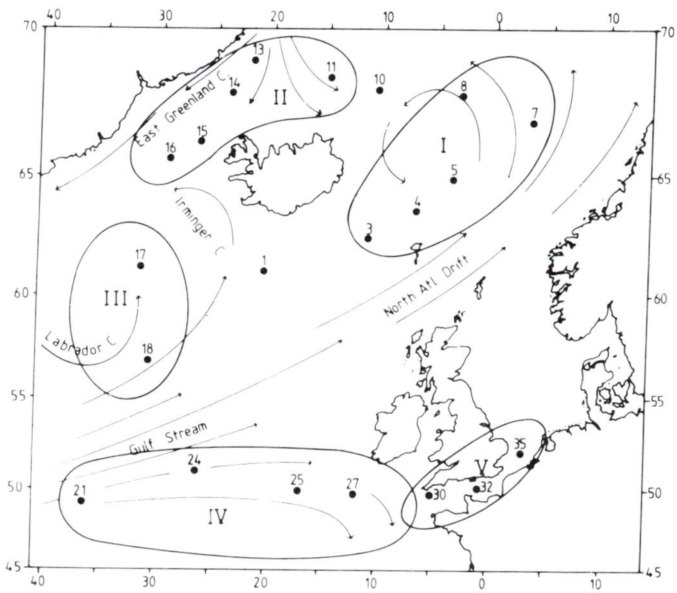

Fig. 4. Schematic representation of surface currents in the north Atlantic Ocean and stations where surface waters were collected. Areas I to V indicate regions with common temperature, salinity and nutrient characteristics.

The conditional stability constants (K') show small differences for these samples. Coastal samples, at lower salinities, tend to have lower K' values (log K' = 7.6) than offshore waters (log K' = 7.8), with again an intermediate value for station 5.

TABLE 6

Complexation capacity for copper (CC), conditional stability constant (K'), salinity and pH of Southern North Sea samples.

Station	CC(nM Cu^{2+})	log K'	S (x 10^3)	pH
1	84	7.6	31.61	7.98
2	80	7.6	32.54	8.21
3	44	7.8	34.71	8.05
4	44	7.8	34.66	7.75
5	64	7.7	33.81	7.93

North Atlantic Ocean. Surface samples in north Atlantic waters were collected at 20 stations in July-August 1983. The area is influenced mainly by the Gulfstream, the North Atlantic Drift and the much colder East Greenland Current. The surface samples can be grouped into fives areas with common characteristics of temperature, salinity, phosphate and silicate concentrations, Fig. 4 (Kramer, 1986).

The CC_{Cu} of the samples ranged 20-66 nM Cu^{2+}, the log K' 7.73 - 8.41. A resemblance can be seen between the different areas (Figs. 5a and 5b). In Table 7 the mean values of the CC_{Cu}, log K' and k_f for the different areas are given. The CC_{Cu}'s of area IV are lower than found for the areas I and II, while an increase can be seen in the Southern Bight of the North Sea, area V. These values compare well with results given before.

Each area is characterized by a practically constant K', but appreciable differences exist between the regions. More stable complexes of copper with natural ligands are formed in the western (III) and northern (I) sections, the stability constants of area IV are the lowest.

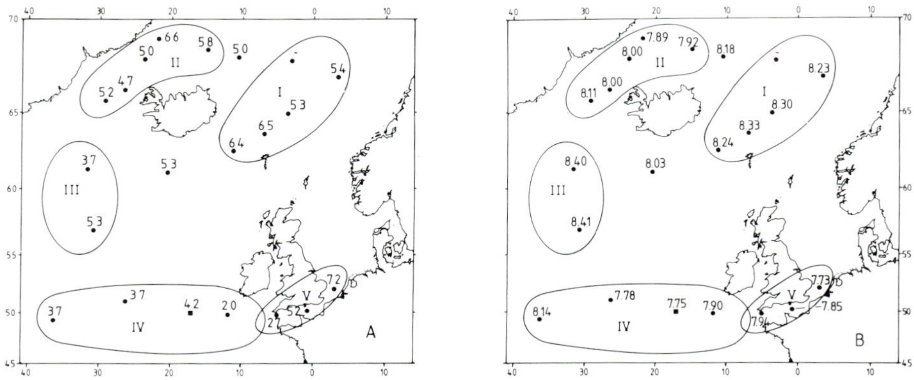

Fig. 5. A) Apparent copper complexation capacity (CC_{Cu}) in nM Cu^{2+} ; B) Conditional stability constants (K') presented as log K'.

TABLE 7

Mean values of complexation capacity (CC), conditional stability constants (K') and rate constants (k_f') for different area in north Atlantic waters.

area	CC(nM Cu^{2+})	log K'	k_f' (x $10^{-4} M^{-1} s^{-1}$)
I	59 ± 6	8.28 ± 0.05	3.6 ± 0.3
II	55 ± 8	7.98 ± 0.09	2.9 ± 0.2
III	45 ± 11	8.41 ± 0.01	-
IV	34 ± 10	7.81 ± 0.08	2.2 ± 0.2
V	-	7.75 ± 0.10	1.4 ± 0.35

The differences between areas can possibly be explained, with several assumptions, in terms of plankton activity patterns. Based on the hypothetical model of Fig. 7, the most important source of organic ligands in these open ocean waters is the in situ production, originating from excretion products and/or intracellular fluids. The amount of ligands produced depends then largely on the biological species present, population density and their activity.

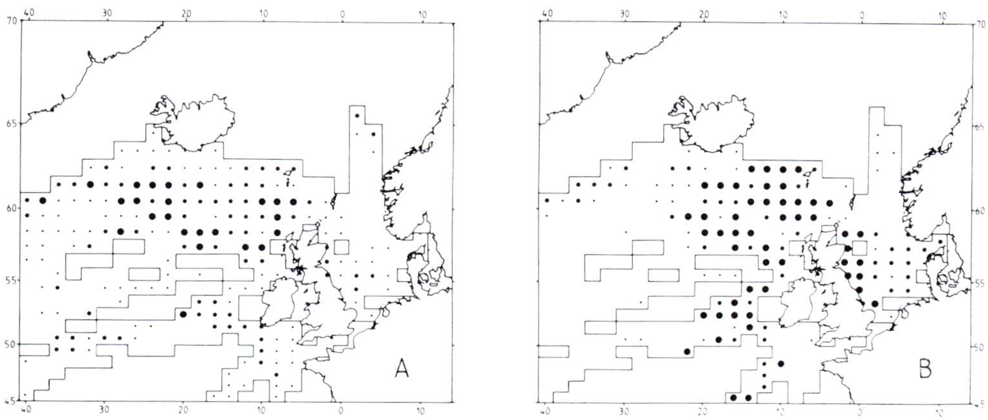

Fig. 6. Simplified representation giving the abbundance and distribution of the 6 major diatom species (A) and 3 major phytoplankton species (B).

The Continuous Plankton Recorder Project (Edinburgh, the Oceanographic Laboratory, 1973) resulted in a Plankton Atlas, covering part of the North Atlantic and the North Sea. Since no data are available for the year of our sampling we have to assume that no drastic changes occur from year to year in the distribution of the organisms, and use the mean data over 10 years of the Plankton Atlas. Some observations with regard to the CC_{Cu} and K' can be made. In Fig. 6a and b the calculated abundance of 6 resp. 3 most abundant diatom and phytoplankton species are presented schematically. Both diatoms and phytoplankton species are most abundant in the area between the UK and Iceland ; an area with very low concentrations of organisms exist in the western and southern part of our sampling areas (III and IV).

It seems from these figures that a positive relation exists between plankton abundance and the complexation capacity. Both exudates and intracellular fluids contain potential ligands, therefore no information on their relative contribution can be obtained from the CC_{Cu}. Differences in K' however might be explained in terms of seasonal variations of plankton activity (Kramer, 1985).

CC_{Cu} across the land-sea boundary. In a transect covering the riverine environment and the open ocean including areas with very different environmental conditions, changes in physical and chemical parameters and changes in the relative sources of complexing agents occur, as presented schematically in Fig. 7. In the riverine and estuarine environment the terrigenuous influence is dominant (both for trace elements and organic matter) ; the resuspension of sediments has an additional influence in shallow coastal waters. In situ biological production and also atmospheric input become progressively, relatively more important towards the open ocean.

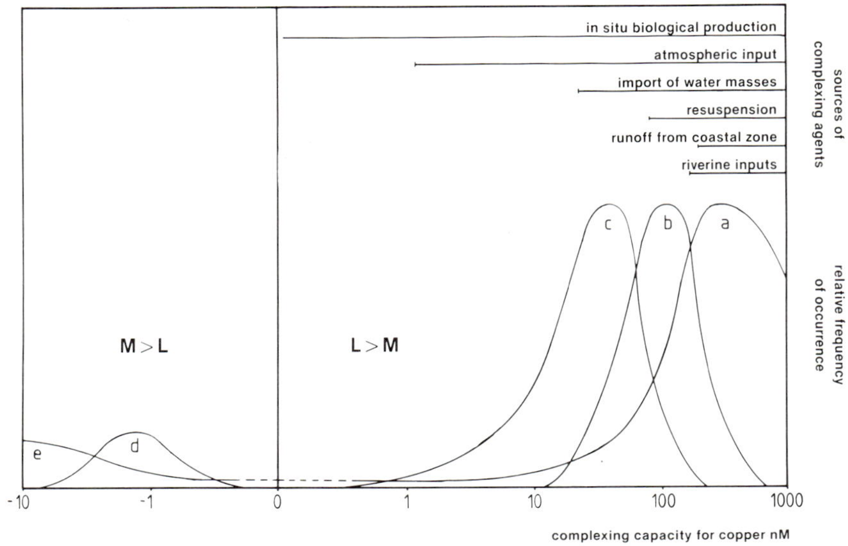

Fig. 7. Schematic model of the distribution of the apparent complexation capacity in river- and estuarine- (a), coastal- (b) and open ocean waters (c). The right part shows the situation L > M (based on Wood et al., 1983), the left part the hypothetical situation M > L for deep oceanic - (d) and estuarine waters (e). Bar graph shows the major sources for each system.

This is reflected in the complexation capacity. Usually a high organic matter content of river and estuarine waters will, together with the colloids in the "dissolved" fraction, result in a high CC_{Cu} (100 - 500 nM Cu^{2+}), Fig. 7a. Coastal waters (Fig. 7b), as a result of mixing with seawater, have a lower CC_{Cu} (60 - 150 nM Cu^{2+}). Open ocean surface waters of the North Atlantic have a CC_{Cu} of 20 - 70 nM Cu^{2+}, which in case of low in situ biological activity might be well below this value (Fig. 7c).

The hypothetical model of Fig. 7 includes two scenarios : the situation found so far in the aquatic environment where L > M, resulting in a positive CC_{Cu}. Theoretically a situation can occur where M > L. One can assume a saturation of the potential ligand sites and/or, after many years of transformations, polycondensations and polymerisations of the organic matter, as one can expect in old deep oceanic waters, a relative decrease in active sites, with (almost) no input of new organic compounds. This can thus result in a "negative" CC_{Cu}, or a situation where the trace metal is partly present in labile uncomplexed form, with no more donor atoms being available for further complexation (Fig. 7d). The limit of this "negative" CC_{Cu} is obviously the total trace metal concentration ; in Fig. 7d, the maximum of the curve being 5 nM Cu actually found in deep oceanic waters (Bruland, 1983). Almost no CC_{Cu} data are available on deep waters. One deep cast in the NE Atlantic shows a decrease in the CC_{Cu} below the euphotic zone towards 22 nM Cu^{2+} at 4820 m (Kramer, 1986).

Theoretically in rivers and estuaries under severe trace metal pollution stress the same situation (M > L) might occur. In this case the limit to the possible "negative" CC_{Cu} is determined by the actual degree of contamination (Fig. 7 e).

The results for the CC_{Cu}, reported in this paper, compare well with the literature data summarized by Kramer (1986) ; most values found for offshore waters differ not more than one order of magnitude, regardless the technique used.

The conditional stability constants found for the different marine areas are not very different. A trend can be observed of K' increasing towards the open ocean. Thus relative stronger complexes are formed in the open ocean than in estuarine and coastal waters. Comparable values have been observed by other authors, using the same technique. Application of different techniques however, can result in several orders of magnitude higher K' (Kramer, 1986). This cannot only be attributed to geographical differences. Another factor is, that different techniques have different sensitivities for the various species of the element concerned.

CONCLUSION

During the last decade many experimental techniques and theoretical models for studies of specific chemical entities of trace elements have been developed or improved. Their application to the natural environment has already resulted in improved understanding of biogeochemical processes in the marine environment, and it is likely that further development in the near future will be significant.

ACKNOWLEDGMENTS

The author is grateful to Dr. J.C. Duinker for the critical review of the manuscript.

REFERENCES

Ahrland, S., 1975. Metal complexes present in seawater. In: The nature of seawater. E.D. Goldberg (ed.), Dahlem Konferenzen, Berlin, pp. 219-244.

Anderson, D.M. and Morel, F.M.M., 1978. Copper sensitivity of Gonyaulax tamarensis. Limnol. Oceanogr., 23: 283-295.

Andreae, M.O., 1977. Determination of arsenic species in natural waters. Anal. Chem., 49: 820-823.

Andreae, M.O., 1979. Arsenic speciation in seawaters : the influence of biological-chemical interactions on the chemistry of a trace element. Limnol. Oceanogr., 24: 440-452.

Andreae, M.O., 1983. The determination of the chemical species of some of the "hydride elements" (arsenic, antimony, tin and germanium) in seawater : methology and results. In: C.S. Wong, E. Boyle, K.W. Bruland, J.D. Burton and E.D. Goldberg (eds). Trace Metals in Seawater. Plenum, New York, pp. 1-19.

Baric, A. and Branica, M., 1967. Polarography of sea water I. Ionic state of state of Cd and Zn in sea water. J. Polarogr. Soc., 13: 4-8.

Batley, G.E. and Florence, T.M., 1976. A novel scheme for the classification of heavy metal species in natural waters. Analyt. Lett., 9: 379-388.

Batley, G.E. and Gardner, D., 1977. Sampling and storage of natural waters for trace metal analysis. Water Res., 11: 745-756.

Benes, P. and Steinnes, E., 1974. In situ dialysis for the determination of the state of trace elements in natural waters. Water Res., 8: 947-953.

Berger, P. and Ewald, M., 1984. Application of the fluorescence quenching titration method to the complexation of copper (II) in the Gironde estuary (France). Mar. Chem., 14: 289-295.

Bewers, J.M. and Windom, H.C., 1982. Comparison of sampling devices for trace metal determinations in seawater. Mar. Chem., 11: 71-86.

Bewers, J.M. and Windom, H.L., 1983. Intercomparison of seawater sampling devices for trace metals. In: C.S. Wong, E. Boyle, K.W. Bruland, J.D. Burton and E.D. Goldberg (eds). Trace Metals in Seawater. Plenum, New York, pp. 143-154.

Bewers, J.M., Yeats, P.A., Westerlund,S., Magnusson, B., Schmidt, D., Zehle, H., Berman, S.S., Mykytiuk, A., Duinker, J.C., Nolting, R.F., Smith, R.G. and Windom, H.L., 1985. An intercomparison of seawater filtration procedures. Mar. Poll. Bull., 16 : 277-281.

Bilinski, H., Huston, R. and Stumm, W., 1976. Determination of the stability constants of some hydroxo and carbonate complexes of Pb(II), Cu(II), Cd(II) and Zn(II) in dilute solutions by anodic stripping voltammetry and differential pulse polarography. Anal. Chim. Acta, 84: 157-164.

Boonlayangoor, C., Allen, H.E. and Noll, K.E., 1979. Colloids in Lake Michigan Water: nature, size distribution, seasonal variation and reaction with metals. Pritzker Dept. of Env. Eng., IIT, Chicago, pp. 213.

Braman, R.S., Johnson, D.L., Foreback, C.C., Ammons, J.M. and Brick, J.C., 1977. Separation and determination of nanogram amounts of inorganic arsenic and methylarsenic compounds. Anal. Chem., 49: 621-625.

Brand, L.E., Sunda, W.G. and Guillard, R.R.L., 1983. Limitation of marine phytoplankton reproductive rates by zinc, manganese and iron. Limnol. Oceanogr., 28: 1182-1198.

Bresnahan, W.T., Grant, C.L. and Weber, J.H., 1978. Stability constants for the complexation of copper (II) ions with water and soil fulvic acids measured by an ion selective electrode. Anal. Chem., 50: 1675-1679.

Brown, S.D. and Kowalski, B.R., 1979. Pseudopolarographic determination of metal complex stability constants in dilute solution by rapid scan ASV. Anal. Chem., 51: 2133-2139.

Bruland, K.W., 1983. Trace elements in seawater. In: J.P. Riley and R. Chester (eds), Chemical Oceanography, Academic Press, London, Vol.8, pp. 157-220.

Bruland, K.W., Franks, R.P., Knauer, G.A. and Martin, J.H., 1979. Sampling and analytical methods for the determination of Cu, Cd, Zn and Ni at the nanogram per liter level in seawater. Anal. Chim. Acta 105: 233-245.

Buffle, J., 1980. A critical comparison of studies of complex formation between copper (II) and fulvic substances of natural waters. Anal. Chim. Acta, 118: 29-44.

Buffle, J., Deladoey, P., Greter, F.L. and Haerdi, W., 1980. Study of the complex formation of copper (II) by humic and fulvic substances. Anal. Chim. Acta, 116 : 255-274.

Buffle, J., Greter, F.L. and Haerdi, W., 1977. Measurement of complexation properties of humic and fulvic acids in natural waters with lead and copper ion selective electrodes. Anal. Chem., 49 : 216-222.

Buffle, J., Teissier, A. and Haerdi, W., 1984. Interpretation of trace metal complexation by aquatic organic matter. In: C.J.M. Kramer and J.C. Duinker (eds), Complexation of Trace Metals in Natural Waters. Nijhoff/Junk Publ., the Hague, pp. 301-316.

Carpenter, E.J. and Livelv, J.S., 1980. Review of estimates of algal growth using ^{14}C tracer techniques. In: P.G. Falkowski (ed.), Primary Productivity in the Sea. Plenum Press, New York, pp. 166-178.

Castilho, P. del, Gerritse, R.G., Marquenie, J.M. and Salomons, W., 1984. Speciation of heavy metals and the in situ accumulation by Dreissena Polymorpha : a new method. In: C.J.M. Kramer and J.C. Duinker (eds), Complexation of Trace Metals in Natural Waters. Nijhoff/Junk Publ., the Hague, pp. 445-448.

Chau, Y.K. and Wong, P.T.S., 1983. Direct speciation analysis of molecular and ionic organometals. In: G.G. Leppard (ed.), Trace Element Speciation in Surface Waters. Plenum, New York, pp. 87-102.

Cleven, R.F.M.J., 1984. Heavy metal/polyacid interaction. An electrochemical study of the binding of Cd(II), Pb(II) and Zn(II) to polycarboxylic and humic acids. Agricultural University, Wageningen, Thesis, pp. 225.

Collier, R. and Edmond, J., 1984. The trace element geochemistry of marine biogenic particulate matter. Prog. Oceanogr., 13: 113-199.

Crosser, M.L. and Allen, H.E., 1977. Determination of complexation capacity of soluble ligands by ion exchange equilibrium. Soil Sci. 123 : 176-181.

Danielson, L.G., 1982. On the use of filters for distinguishing between dissolved and particulate fractions in natural waters. Water Res., 16: 179-182.

Degens, E.T. and Ittekot, V.I., 1983. Dissolved organic carbon- an overview. Mitt. Geol. Palaeont. Inst. Univ. Hamburg 55: 21-38.

Duinker, J.C., 1980. Suspended matter in estuaries : adsorption and desorption processes. In: E. Olausson and I. Cato (eds), Chemistry and Biogeochemistry of Estuaries. Wiley, New York, pp. 121-151.

Duinker, J.C. and Kramer, C.J.M., 1977. An experimental study on the speciation of dissolved Zn, Cd, Pb and Cu in river Rhine and North Seawater, by DPASV, Mar. Chem. 5: 207-228.

Duinker, J.C. and Nolting, R.F., 1977. Dissolved and particulate trace metals in the Rhine estuary and the southern Bight. Mar. Poll. Bull., 8: 65-71.

Duinker, J.C., Nolting, R.F. and van der Sloot, H.A., 1979. The determination of suspended metals in coastal waters by different sampling and processing techniques (filtration, centrifugation). Neth. J. Sea Res., 13: 282-297.

Duursma, E.K. and Dawson, R. (eds), 1982. Marine organic chemistry. Elsevier, Amsterdam, pp. 521.

Dyrssen, D. and Wedborg, M., 1974. Equilibrium calculations of the speciation of elements in seawater. In: E.D. Goldberg (ed.), The Sea, Vol. V, Marine Chemistry . Wiley, pp. 181-195.

Dyrssen, D. and Wedborg, M., 1980. Chemical speciation of some constituents in estuarine waters. In: E. Olausson and I. Cato (eds), Chemistry and Biogeochemistry of Estuaries. Wiley, New York, pp. 71-119.

Edinburgh, Oceanographic Laboratory, 1973. Continuous plankton records : a plankton atlas of the North Atlantic and the North Sea. Bull. Mar. Ecol., 7: 1-74.

Elderfield, H., 1970. Chromium speciation in seawater. Earth Planet. Sci. Lett., 9: 10-16.

Ernst, R., Allen, H.E. and Mancy, K.H., 1975. Characterization of trace metal species and measurement of trace metal stability constants by electrochemical techniques. Water Res., 9: 969-979.

Evans, G., Johnson, P.E., Brushmiller and Ames, R.W., 1979. Detection of labile zinc-binding ligands in biological fluids by modified gel filtration chromatography. Anal. Chem., 51: 839-843.

Fisher, N.S. and Fabris, J.G., 1982. Complexation of Cu, Zn and Cd by metabolites excreted from marine diatoms. Mar. Chem., 11: 245-255.

Fitzwater, S.E., Knauer, G.A. and Martin, J.H., 1982. Metal contamination and its effect on primary production measurements. Limnol. Oceanogr., 27: 544-551.

Florence, T.M. and Batley, G.E., 1980. Chemical speciation in natural waters. C.R.C. Crit. Rev. Anal. Chem., 9: 219-296.

Gamble, D.S., Langford, C.H. and Underdown, A.W., 1984. The interrelationship of aggregation and cation binding of fulvic acid. In: C.J.M. Kramer and J.C. Duinker (eds), Complexation of Trace Metals in Natural Waters. Nijhoff/Junk Publ., the Hague, pp. 349-356.

Gamble, D.S. and Schnitzer, M., 1973. The chemistry of fulvic acid and its reactions with metal ions. In: Trace metals and metal-organic interactions in natural waters. Ann Arbor Sci. Publ. pp. 265-302.

Gardner, W.S., Landrum, P.F. and Yates, D.A., 1982. Fractionation of metal forms in natural waters by size exclusion chromatography with inductively coupled argon plasma detection. Anal. Chem., 54: 1196-1198.

Ghassemi,M., Christman, R.F., 1968. Properties of the yellow organic acids of natural waters. Limnol. Oceanogr., 13: 583-597.

Gillespie, P.A. and Vaccaro, R.F., 1978. A bacterial bioassay for measuring the copper chelation capacity of seawater. Limnol. Oceanogr., 23: 543-548.

Gnassia-Barelli,M., Romeo, M., Laumond, F. and Pesando, D., 1978. Experimental studies on the relationship between natural copper complexes and their toxicity to phytoplankton. Mar. Biol., 47: 15-19.

Gould, J.P., 1982. The kinetics of hexavalent chromium reduction by metallic iron. Water Res., 16: 871-878.

Guy, R.D. and Chakrabarti, C.L., 1976. Studies of metal-organic interactions in model systems pertaining to natural waters. Can. J. Chem., 54: 2600-2611.

Harrison, R.M. and Laxen, D.P.H., 1980. Physicochemical speciation of lead in drinking water. Nature, 286: 791-793.

Hart, B.T., 1981. Trace metal complexing capacity of natural waters : a review. Environm. Technol. Lett., 3: 95-110.

Hart, B.T. and Davies, S.H.R., 1977. A new dialysis-ion exchange technique for determining the forms of trace metals in water. Aust. J. Mar. Freshw. Res., 28: 105-112.

Hart, B.T. and Davies, S.H.R., 1981. Trace metal speciation in the freshwater and estuarine regions of the Yarra river, Victoria. Est. Coast. Shelf Sci., 12: 353-374.

Hasle, J.R. and Abdullah, M.I., 1981. Analytical fractionation of dissolved copper, lead and cadmium in coastal seawater. Mar. Chem., 10: 487-503.

Hirose, K., Dokiya,Y. and Sugimura, Y., 1982. Determination of conditional stability constants of organic copper and zinc complexes dissolved in seawater using ligand exchange method with EDTA. Mar. Chem., 11: 343-354.

Hoffman, M.R., 1981. Thermodynamic, kinetic and extrathermodynamic considerations in the development of equilibrium models for aquatic systems. Environm. Sci. Technol., 15: 345-353.

Hunt, C.D. and Fitzgerald, W.F., 1983. The capacity of marine plankton, macrophytes and particulate matter to adsorb Cu^{2+} in the presence of Mg^{2+}. Mar. Chem., 12: 255-280.

Hunter, K.A. and Liss, P.S., 1982. Organic matter and the surface charge of suspended particles in estuarine waters. Limnol. Oceanogr., 27: 322-335.

Imber, B.E. and Robinson, M.G., 1983. Complexation of zinc by exudates of Thalassiosira fluviatilis grown in culture. Mar. Chem., 14: 31-41.

Imber, B.E., Robinson, M.G. and Pollehne, F., 1984. Complexation by diatom exudates in culture and in the field. In: C.J.M. Kramer and J.C. Duinker (eds), Complexation of Trace Metals in Natural Waters. Nijhoff/Junk Publ., the Hague, pp. 431-442.

Jackson, G.A. and Morgan, J.J., 1978. Trace metal-chelator interactions and phytoplankton growth in seawater media. Limnol. Oceanogr., 23: 268-282.

Jagner, D. and Aren, K., 1982. Application of PSA to complexiometric titrations. Anal. Chim. Acta, 134: 201-209.

Jardim, W.F. and Allen, H.E., 1984. Measurement of copper complexation by naturally occurring ligands. In: C.J.M. Kramer and J.C. Duinker (eds), Complexation of Trace Metals in Natural Waters. Nijhoff/Junk Publ., the Hague, pp. 1-15.

Jenne, E.A. (ed.), 1979. Chemical modelling in aqueous systems : speciation, sorption, solubility and kinetics. ACS Symp. series 93. pp. 1-194.

Johnston, R., 1964. Seawater, the natural medium of phytoplankton. II. Trace metals and chelation, and general discussion. J. Mar. Biol. Assoc. U.K., 44: 87-109.

Kennedy, V.C., Zellweger, G.W. and Jones, B.F., 1974. Filter pore-size effects on the analysis of Al, Fe, Mn and Ti in water. Water Resourc. Res., 10: 785-790.

Kerr, R.A. and Quinn, J.G., 1980. Chemical comparison of dissolved organic matter isolated from different oceanic environments. Mar. Chem., 8: 217-229.

Kester, D.R. (rapporteur), 1975. Chemical speciation in seawater. In: E.D. Goldberg (ed.), The Nature of Seawater. Dahlem Conf., Berlin, pp. 17-41.

Khailov, K.M., 1964. The formation of organic metallic complexes with the participation of external metabolites of marine algae. Dokl. Adad. Nauk. SSSR. Biol. Sci., 155: 237-239.

Kramer, C.J.M., 1985. On the copper complexation capacity in the marine environment. Thesis. University of Groningen, The Netherlands, pp. 133.

Kramer, C.J.M., 1986. Apparent copper complexation capacity and conditional stability constants in north atlantic waters and their relation to plankton activity. Mar. Chem. 17 : (in press).

Kramer, C.J.M. and Duinker, J.C., 1984a. Complexation capacity and conditional stability constants for copper of sea- and estuarine waters, sediment extracts and colloids. In: C.J.M. Kramer and J.C. Duinker (eds), Complexation of Trace Metals in Natural Waters. Nijhoff/Junk Publ., the Hague, pp. 217-228.

Kramer, C.J.M. and Duinker, J.C. (eds), 1984b. Complexation of trace metals in natural waters. Proceedings of the symposium, Texel, May 2-6 1983. Nijhoff/Junk Publ. The Hague, pp. 448.

Kramer, C.J.M. and Manshanden, G.M., 1982. Improvements in automatic electrochemical analyses,using the PAR 374. NIOZ report, 1982, 4.

Krom, M.D. and Sholkovitz, E.R., 1978. On the association of iron and manganese with organic matter in anoxic marine pore waters. Geochim. Cosmochim. Acta, 42: 607-611.

Laane, R.W.P.M., 1982. Chemical characteristics of the organic matter in the water phase of the Ems-Dollart estuary. Thesis, University of Groningen, 134 pp.

Laane, R.W.P.M. and Kramer, C.J.M., 1984. Complexation of Cu^{2+} with humic substances in relation to different extraction procedures of sandy and silty marine sediments. In: C.J.M. Kramer and J.C. Duinker (eds), Complexation of Trace Metals in Natural Waters. Nijhoff/Junk Publ., the Hague, pp. 345-348.

Lammers, W.T., 1967. Separation of suspended and colloidal particles from natural water. Environm. Sci. Technol., 1: 52-57.

Langston, W.J. and Bryan,G.W., 1984. The relationships between metal speciation in the environment and bioaccumulation in aquatic organisms. In: C.J.M. Kramer and J.C. Duinker (eds), Complexation of Trace Metals in Natural Waters. Nijhoff/Junk Publ., the Hague, pp. 375-392.

Laxen, D.P.H. and Chandler, I.M., 1982. Comparison of filtration techniques for size distribution in fresh waters. Anal. Chem., 54: 1350-1355.

Lee, J., 1983. Complexation analysis of fresh waters by equilibration diafiltration. Water Res., 17: 501-510.

Leeuwen, H.P. van, Threels, W.F. and Cleven, R.F.M.J., 1981. Pulse polarography of heavy metal ions in polyelectrolyte solutions : behaviour of Pb^{2+} and Cd^{2+} in PMA solutions. Coll. Czech Chem. Commun., 46: 3027-3037.

Leppard, G.G., (ed.), 1983. Trace element speciation in surface waters and its ecological implications. Plenum Press. New York, 320 pp.

Mackey, D.J., 1982. An investigation of the suitability of amberlite XAD-1 resin for studying trace metal speciation in seawater. Mar. Chem., 11: 169-181.

Mackey, D.J., 1983a. The strong complexing capacity of seawater-an investigation of south eastern Australian coastal waters. Mar. Chem., 14: 73-87.

Mackey, D.J., 1983b. Metal-organic complexes in seawater- an investigation of naturally occurring complexes of Cu, Zn, Fe, Mg, Ni, Cr, Mn and Cd using high performance liquid chromatography with atomic fluorescence detection. Mar. Chem., 13: 169-180.

Magjer, T. and Branica, M., 1977. A new electrode system with efficient mixing of electrolyte. Croatica Chem. Acta, 49: L1-L5.

Mantoura, R.F.C., 1982. Organo-metallic interactions in natural waters. In: E.K. Duursma and R. Dawson (eds). Marine organic chemistry Elsevier, Amsterdam, pp. 179-223.

Mantoura, R.F.C., Dickson, A. and Riley, J.P., 1978a. The speciation of trace metals with humic compounds in natural waters. Thalassia Jugosl., 14: 127-145.

Mantoura, R.F.C., Dickson, A. and Riley, J.P., 1978b. The complexation of metals with humic materials in natural waters. Est. Coast. Mar. Sci., 6: 387-408.

Mantoura, R.F.C. and Riley, J.P., 1975. The analytical concentration of humic substances from natural waters. Anal. Chim. Acta, 76: 97-106.

Mart, L., 1979a. Prevention of contamination and other accuracy risks in voltammetric trace metal analysis of natural waters. Fres. Z. Anal. Chem., 296: 350-357.

Mart, L., 1979b. Prevention of contamination and other accuracy risks in voltammetric trace metal analysis of natural waters. 2. Collection of surface water samples. Fres. Z. Anal. Chem., 299: 97-102.

Mayer, L.M., 1982. Aggregation of colloidal ion during estuarine mixing : kinetics, mechanism and seasonality. Geochim. Cosmochim. Acta, 46: 2527-2535.

McCrady, J.K. and Chapman, G.A., 1979. Determination of copper complexing capacity of natural river water, well water and artificially reconstituted water. Water Res., 13 : 143-150.

McKnight, D.M. and Morel, F.M.M., 1979. Release of weak and strong copper complexing agents by algae. Limnol. Oceanogr., 24: 823-837.

Mill, A.J.B., 1980. Colloidal and macromolecular forms of iron in natural waters. 2. Their occurrence in rivers of south west England. Environm. Technol. Lett., 1: 109-124.

Mills, G.L., Hanson, A.K., Quinn, J.G. and Lammela, W.L., 1982. Chemical studies of copper-organic complexes isolated from estuarine waters using C_{18} reverse-phase liquid chromatography. Mar. Chem., 11: 355-377.

Mills, G.L. and Quinn, J.G., 1984. Dissolved copper and copper-organic complexes in the Narragansett Bay estuary. Mar. Chem., 15: 151-172.

Moffett, J.W. and Zika, R.G., 1983. Oxidation kinetics of Cu(I) in seawater : implications for its existence in the marine environment. Mar. Chem., 13: 239-251.

Moore, R.M., Burton, J.D., Williams, P.J., Le B. and Young, M.L., 1979. The behaviour of dissolved organic material, iron and manganese in estuarine mixing. Geochim. Cosmochim. Acta, 43: 919-926.

Morel, F.M.M. and Morel-Laurens, N.M.L., 1983. Trace metals and plankton in the oceans:

facts and speculations. In: C.S. Wong, E. Boyle, K.W. Bruland, J.D. Burton and E.D. Goldberg (eds). Trace Metals in Seawater. Plenum, New York, pp. 841-869.

Morgan, J.J., 1966. Chemical equilibria and kinetic properties of manganese in natural waters. In: S.D. Faust and J. Hunter (eds), Principle and Applications of Water Chemistry. Wiley, New York.

Nakayama, E., Kuwamoto, T., Surubo, S.T. and Fujinaga, T., 1981. Chemical speciation of chromium in seawater. Anal. Chim. Acta, 30: 401-404.

Neihof, R.H. and Loeb, G.I., 1972. The surface charge of particulate matter in seawater. Limnol. Oceanogr., 17: 7-16.

Neubecker, T.A. and Allen, H.E., 1983. The measurement of complexation capacity and conditional stability constants for ligands in natural waters. Water Res., 17: 1-14.

Nordstrom, D.K. and Ball, J.W., 1984. Chemical models, computer programs and metal complexation in natural waters. In: C.J.M. Kramer and J.C. Duinker (eds), Complexation of Trace Metals in Natural Waters. Nijhoff/Junk Publ., the Hague, pp. 149-164.

Nürnberg, H.W., 1983. Voltammetric studies on trace metal speciation in natural waters. Part II Application and conclusions for chemical oceanography and chemical limnology. In: G.G. Leppard (ed.), Trace Element Speciation in Surface Waters. Plenum, New York, pp. 211-230.

Nürnberg, H.W., 1984. Potentialities of voltammetry in the study of physicochemical aspects of heavy metal complexation in natural waters. In: C.J.M. Kramer and J.C. Duinker (eds), Complexation of Trace Metals in Natural Waters. Nijhoff/Junk Publ., the Hague, pp. 95-115.

Oda, H. and Okabe, S., 1977. Some speciations of zinc and copper in coastal and oceanic waters by ultrafiltration. J. Fac. Mar. Sci. Technol., Tokai Univ., 10: 81-90.

Pankow, J.F. and Morgan, J.J., 1981a. Kinetics for the aquatic environment. Environm. Sci. Technol., 15: 1155-1164.

Pankow, J.F. and Morgan, J.J., 1981b. Kinetics for the aquatic environment. Part II. Environm. Sci. Technol., 15: 1306-1313.

Parks, G.A., 1975. Adsorption in the marine environment. In: J.P. Riley and G. Skirrow (eds), Chemical Oceanography. Academic Press, London, Vol.1, 2nd ed., pp. 241-308.

Paulsen, A.J., 1978. Potentiometric studies of cupric hydroxide complexation. Thesis. Graduate School of Oceanography, Univ. of Rhode Island USA.

Plavsic, M. and Branica, M., 1985. Differentiation of substances yielding Cu complexing capacity in seawater by SEP-PAK C_{18}. Mar. Chem. (in press).

Plavsic, M., Kozar, S., Krznaric, D., Bilinski, H. and Branica, M., 1980. The influence of organics on the adsorption of copper (II) on γ-Al_2O_3 in seawater, model studies with EDTA. Mar. Chem., 9: 175-182.

Plavsic, M., Krznaric, D. and Branica, M., 1982. Determination of the apparent copper complexing capacity of seawater by ASV. Mar. Chem., 11: 17-31.

Ragan, M.A., Smidsröd, Larsen, B., 1979. Chelation of divalent metal ions by brown algal polyphenols. Mar. Chem., 7: 265-271.

Raspor, B., 1980. Distribution and speciation of cadmium in natural waters. In: J.O.

Nriagu (ed.), Cadmium in the Environment. Wiley, New York, Part. 1, pp. 147-236.

Raspor, B. and Branica, M., 1975. Comparison of the polarographic reduction of Cd-NTA, -EDTA and -DTPA chelates in chloride solutions. J. Electroanal. Chem., 60: 335-339.

Raspor, B., Nürnberg, H.W., Valenta, P. and Branica, M., 1980. Kinetics and mechanism of trace metal chelation in seawater. J. Electroanal. Chem., 115: 293-308.

Raspor, B., Nürnberg, H.W., Valenta, P. and Branica, M., 1984. Studies in seawater and lakewater on interactions of trace metals with humic substances isolated from marine and estuarine sediments. II. Voltammetric investigations on trace metal complex formation in the dissolved phase. Mar. Chem., 15: 231-249.

Raspor, B., Valenta, P., Nürnberg, H.W. and Branica, M., 1977. Polarographic studies on the kinetics and mechanisms of Cd(II)- chelate formation with EDTA in seawater. Thalassia Jugosl., 13: 79-91.

Raspor, B., Valenta, P., Nürnberg, H.W. and Branica, M., 1978. The chelation of cadmium with NTA in seawater as a model for the typical behaviour of trace heavy metal chelates in natural waters. Sci. Total Environm., 9: 87-109.

Reuter, J.H. and Perdue, E.M., 1977. Importance of heavy metal - organic matter interactions in natural waters. Geochim. Cosmochim. Acta, 41: 325-334.

Roekens, E.J. and Van Grieken, R.E., 1983. Kinetics of iron (II) oxidation in seawater of various pH. Mar. Chem., 13: 195-202.

Röhl, R., 1982. Trace metal speciation in seawater - a paper electrophoretic approach. Anal. Chim. Acta, 135: 99-110.

Ruzic, I., 1982. Theoretical aspects of the direct titration of natural waters and its information yield for trace metal speciation. Anal. Chim. Acta, 140: 99-113.

Ruzic, I. and Nikolic, S., 1982. The influence of kinetics on the direct titration curves of natural water systems - theoretical considerations. Anal. Chim. Acta, 140: 331-334.

Ryan, D.K. and Weber, J.H., 1982. Fluorescence quenching titration for determination of complexing capacities and stability constants of fulvic acids. Anal. Chem., 54: 986-990.

Salbu, B., 1981. New trends in analysis of ground water. Mikrochim. Acta (Wien) II: 351-359.

Schnitzer, M., 1976. The chemistry of humic substances. In: J.O. Nriagu (ed.). Environmental biogeochemistry, vol. 1. Ann. Arbor Sci., p. 89-108.

Serritti, A., Pellegrini, D., Barhigiani, C. and Ferrara, R., 1985. Copper complexing capacity of phytoplanktonic cell exudates. Mar. Chem. (submitted).

Seritti, A., Pellegrini, D., Ferrara, R. and Barhigiani, C., 1983. Presence and speciation of Zn, Cd, Pb and Cu in the Tyrrhenian Sea. In: Proceedings of the conference "Heavy metals in the environment", Heidelberg, 2: 1120-1123.

Shapiro, J., 1964. Effect of yellow organic acids on iron and other metals in water. J. Am. Waste W. Ass., 56: 1062-1082.

Sholkovitz, E.R., 1976. Flocculation of dissolved organic and inorganic matter during mixing of river water and sea water. Geochim. Cosmochim. Acta, 40: 834-845.

Sholkovitz, E.R., 1978. The flocculation of dissolved Fe, Mn, Al, Cu, Ni, Co and Cd during estuarine mixing. Earth Planet. Sci. Lett., 41: 77-86.

Shuman, M.S. and Michael, L.C., 1975. Ring-disk techniques for studying kinetics of metal complex formation and dissociation in natural waters. In: proceedings Int. Conf. "Heavy metals in the environment", vol. 1, Toronto, Canada, p. 227.

Shuman, M.S. and Michael, L.C., 1978. Application of the rotated disk electrode to measurement of copper complex dissociation rate constants in marine coastal samples. Environm. Sci. Technol., 12: 1069-1072.

Sibley, T.H. and Morgan, J.J., 1975. Equilibrium speciation of trace metals in freshwater - seawater mixtures. In: T. Hutchinson (ed.), Proc. Int. Conf. Heavy Metals in the Environment. Univ. of Toronto, Ontario, p. 319.

Sigg, L., Stumm, W. and Zinder, B., 1984. Chemical processes at the particle-water interface ; implications concerning the form of occurrence of solute and adsorbed species. In: C.J.M. Kramer and J.C. Duinker (eds), Complexation of Trace Metals in Natural Waters. Nijhoff/Junk Publ., the Hague, pp. 251-266.

Sigleo, A.C. and Helz, G.R., 1981. Composition of estuarine colloidal material: major and trace elements. Geochim. Cosmochim. Acta, 45: 2501-2509.

Sigleo, A.C., Hoering, T.C. and Helz, G.R., 1982. Composition of estuarine colloidal material: organic components. Geochim. Cosmochim. Acta, 46: 1619-1626.

Sillen, L.G. and Martell, A.E., 1964, 1971. Stability constants of metal-ion complexes. Chem. Soc. Spec. Publ., n° 17, n° 25.

Simoes-Goncalves, M.L.S. and Valenta, P., 1982. Voltammetric and potentiometric investigations on the complexation of Zn(II) by glycine in seawater. J. Electroanal. Chem., 132: 357-375.

Simoes-Goncalves, M.L.S., Vaz, M.C.T.A. and Frausto da Silva, J.J.R., 1981. Stability constants of chloro-complexes of Cd(II) in seawater medium. Talanta, 28: 237-240.

Sipos, L., Valenta, P., Nürnberg, H.W. and Branica, M., 1980. Voltammetric determination of the stability constants of the predominant labile lead complexes in seawater. In: M. Branica and Z. Konrad (eds), Lead in the Marine Environment. Pergamon, Oxford, pp. 61-76.

Smith, R.G., 1976. Evaluation of combined applications of ultrafiltration and complexation capacity techniques to natural waters. Anal. Chem., 48: 74-76.

Spencer, M.J., Piotrowicz, S.R. and Betzer, P.R., 1982. Concentrations of cadmium, copper, lead and zinc in surface waters of the northwest Atlantic Ocean - a comparison of Go-Flo and teflon water samplers. Mar. Chem., 11: 403-410.

Srna, R.F., Garrett, K.S., Miller, S.M. and Thum, A.B., 1980. Copper complexation capacity of marine water samples from southern California. Environm. Sci. Technol., 14: 1482-1486.

Stiff, M.J., 1971. The chemical states of copper in polluted fresh water and a scheme of analysis to differentiate them. Water Res., 5: 585-599.

Stolzberg, R.J., 1977. Potential inaccuracy in trace metal speciation measurements by DPP. Anal. Chim. Acta, 92: 193-196.

Stolzberg, R.J. and Rosin, D., 1977. Chromatographic measurement of submicromolar strong complexing capacity in phytoplankton media. Anal. Chem., 49: 225-230.

Stuermer, D.H. and Harvey, G.R., 1977. The isolation of humic susbtances and alcohol-soluble organic matter from seawater. Deep Sea Res., 24: 303-309.

Stumm, W. and Bilinski, H., 1972. Trace metals in natural waters: difficulties of interpretation arising from our ignorance on their speciation. In: S.H. Jenkins (ed.), Advances on Water Pollution Research. Pergamon, New York, pp. 39-52.

Stumm, W. and Brauner, P.A., 1975. Chemical speciation. In: J.P. Riley and G. Skirrow (eds). Chemical Oceanography, vol. I, 2nd ed. Academic Press, New York, p. 173-234.

Stumm, W. and Lee, G.F., 1961. Oxygenation of ferrous iron. Ind. Eng. Chem., 53: 143-146.

Stumm, W. and Morgan, J.J., 1981. Aquatic chemistry 2nd. ed., Wiley Interscience, New York, 780 pp.

Sueur, S., van den Berg, C.M.G. and Riley, J.P., 1982. Measurement of the metal complexing ability of exudates of marine macroalgae. Limnol. Oceanogr., 27: 536-543.

Sugai, S.F. and Healy, M.L., 1978. Voltammetric studies of the organic association of copper and lead in two Canadian inlets. Mar. Chem., 6: 291-308.

Sugimura, Y., Suzuki, Y. and Miyake, Y., 1978. Chemical forms of minor metallic elements in the ocean. J. Oceanogr. Soc. Japan, 34: 93-96.

Sunda, W.G. and Ferguson, R.L., 1983. Sensitivity of natural bacterial communities to additions of copper and to cupric ion activity : a bioassay of copper complexation in seawater. In: C.S. Wong, E. Boyle, K.W. Bruland, J.D. Burton and E.D. Goldberg (eds). Trace Metals in Seawater. Plenum, New York, pp. 871-891.

Sunda, W. and Guillard, R.R.L., 1976. The relationship between cupric ion activity and the toxicity of copper to phytoplankton. J. Mar. Res., 34: 511-529.

Sung, W. and Morgan, J.J., 1980. Kinetics and product of ferrous iron oxygenation in aqueous systems. Environm. Sci. Technol., 14: 561-568.

Swallow, K.C., Westall,J.C., McKnight, D.S., Morel, N.M.L. and Morel, F.M.M., 1978. Potentiometric determination of copper complexation by phytoplankton exudates. Limnol. Oceanogr., 23: 538-542.

Sylva, R.N. and Florence, T.M., 1980. Cited In: T.M. Florence and G.E. Batley CRC Rev. Anal. Chem., 9: 219-296.

Takamatsu, T. and Yoshida, T., 1978. Determination of stability constants of metal-humic-acid complexes by potentiometric titration and ion selective electrodes. Soil Sci., 125: 377-386.

Thurman, E.M., Wershaw, R.L., Malcolm, R.L. and Pinckney, D.J., 1982. Molecular size of aquatic humic substances. Org. Geochem., 4: 27-35.

Truitt, R.E. and Weber, J.H., 1981. Copper II - and cadmium II - binding abilities of some New Hampshire freshwaters determined by dialysis titration. Environm. Sci. Technol., 15: 1204-1208.

Underdown, A.W., Langford, C.H. and Gamble, D.S., 1981. Light scattering of a polydisperse Fulvic Acid. Anal. Chem., 53: 2139-2140.

Valenta, P., 1983. Voltammetric studies on trace metal speciation in natural waters. I. Methods. In: G.G. Leppard (ed.), Trace Element Speciation in Surface Waters. Plenum, New York, pp. 49-67.

Van den Berg, C.M.G., 1982. Determination of copper complexation with natural organic ligands in seawater by equilibration with MnO_2. II. Experimental procedures and application to surface seawater. Mar. Chem., 11: 323-342.

Van den Berg, C.M.G., 1983a. Organic and inorganic speciation of copper in the Irish Sea. Mar. Chem., 14: 201-212.

Van den Berg, C.M.G., 1983b. Trace metal speciation in seawater. Anal. Proc., 20: 458-460.

Van den Berg, C.M.G., 1984. Determination of the complexing capacity and conditional stability constants of complexes of copper (II) with natural organic ligands in seawater by CSV of copper-catechol complex ions. Mar. Chem., 15: 1-18.

Van den Berg, C.M.G. and Kramer, J.R., 1979. Determination of complexing capacities of ligands in natural waters and conditional stability constants of the copper complexes by means of manganese dioxide. Anal. Chim. Acta, 106: 113-120.

Van der Weijden, C.H. and Reith, M., 1982. Chromium III - Chromium VI interconversions in seawater. Mar. Chem., 11: 565-572.

Wallace, G.T., 1982. The association of copper, mercury and lead with surface-active organic matter in coastal seawater. Mar. Chem., 11: 379-394.

Weber, J.H. and Wilson, S.A., 1975. The isolation and characterisation of FA and HA from riverwater. Water Res., 9: 1079-1084.

Wedborg, M., 1979. Studies on complexation in seawater. Thesis. University of Göteborg, Sweden.

Whitfield, M. and Turner, D.R., 1980. The theorerical studies of the chemical speciation of lead in seawater. In: M. Branica and Z. Konrad (eds), Lead in the Marine Environment. Pergamon, Oxford, pp. 109-148.

Wood, A.M., Evans, D.W. and Alberts, J.J., 1983. Use of an ion exchange technique to measure copper complexation capacity on the continental shelf of the southeastern United States and in the Sargasso Sea. Mar. Chem., 13: 305-326.

Zika, R.G., 1982. Marine organic photochemistry. In: E.K. Duursma and R. Dawson (eds). Marine organic chemistry. Elsevier Amsterdam, pp. 299-325.

Zirino, A., and Yamamoto, S., 1972. A pH-dependent model for the chemical speciation of Cu, Zn, Cd and Pb in seawater. Limnol. Oceanogr., 17: 661-671.

Zuehlke, R.W. and Kester, D.R., 1983. Copper speciation in marine waters. In: C.S. Wong, E. Boyle, K.W. Bruland, J.D. Burton and E.D. Goldberg (eds). Trace Metals in Seawater. Plenum, New York, pp. 773-788.

Zutic, V., Cosovic, B., Marcenko, E. and Bihari, N., 1981. Surfactant production by marine phytoplankton. Mar. Chem., 10: 505-520.

RIVER INPUTS INTO OCEANS

W. MICHAELIS, V. ITTEKKOT and E.T. DEGENS
Geologisch-Paläontologisches Institut, Bundesstrasse 55, D-2000 Hamburg 13 (Federal Republic of Germany)

ABSTRACT

Michaelis, W., Ittekkot, V. and Degens, E.T., 1986. River inputs into oceans. In: P. Lasserre and J.M. Martin (eds), Biogeochemical Processes at the Land-Sea Boundary. Elsevier, Amsterdam.

Data collected during an ongoing SCOPE/UNEP study on carbon transport in major world rivers are presented. The annual global transport rates are 0.42 x 10^9 t C/a for POC and 0.11 x 10^9 t C/a for DOC. Strong seasonal fluctuations were observed in POC and DOC transport, indicating the significance of continuous monthly sampling programs.

INTRODUCTION

The rivers play a major role in the transfert of carbon and mineral nutrients from land to the sea and influence significantly the biogeochemical processes operating in coastal waters. Quantification of the material transport, both in the dissolved and particulate forms, has been attempted by several authors in the past (Clarke, 1924 ; Holeman, 1968 ; Garrels & McKenzie, 1971 ; Martin et al., 1980 ; Meybeck, 1982 ; Milliman & Meade, 1983). Depending on the type of sampling techniques and methods of calculations employed there are differences in the reported fluxes. A major problem in such calculations is the paucity of reliable data from some of the major rivers of the world especially of Asia (see e.g. Milliman & Meade, 1983). Additionally the difficulty of obtaining representative samples from the rivers will adversely affect flux calculations. Most of the inferences drawn on the nature and transport of riverine materials rest on data collected randomly - at different points in time and space. Seasonal variations in the transport of materials are very common in some of the major world rivers, and in some cases more than 60 % of the material transport occurs within a very short period of time. Furthermore, available data are not always comparable since the analytical techniques used differ from river to river.

To overcome some of these difficulties we started a program to monitor the inorganic and organic constituents in major world rivers. The study "Carbon Transport in Major World Rivers" is sponsored by the Scientific Committee for Problems of the Environment (SCOPE) and the United Nations Environment Program (UNEP). Within its framework over 50 % of the total water discharged to the sea is being monitored at least once a month.

The major rivers investigated are given in Fig. 1. In this article we report on the nature and content of organic carbon carried by the major world rivers based on data collected during the first two years of the study. Some data on studies on the european estuaries are also presented. For sampling locations and basic data see the chapters on individual rivers in Degens (1982), Degens et al. (1983, 1985).

Fig. 1. Rivers being studied during the SCOPE/UNEP Program.

POOLS OF ORGANIC CARBON

The major pools of organic carbon on land and in the sea are depicted in Fig. 2a and b. Soil organic carbon is by far the most substantial pool followed by dissolved organic

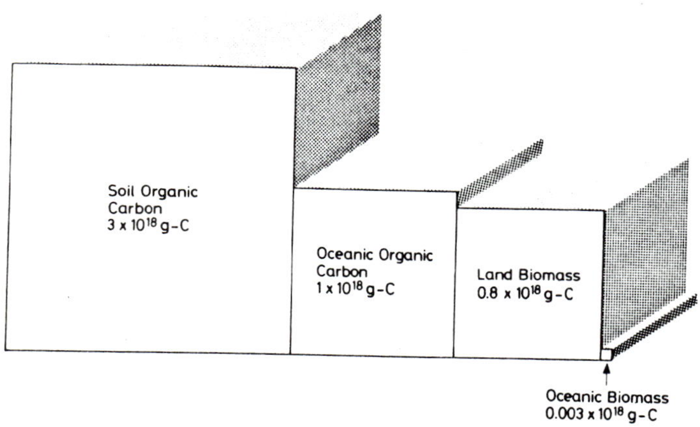

Fig. 2a. The major pools of organic carbon on land and in the sea.

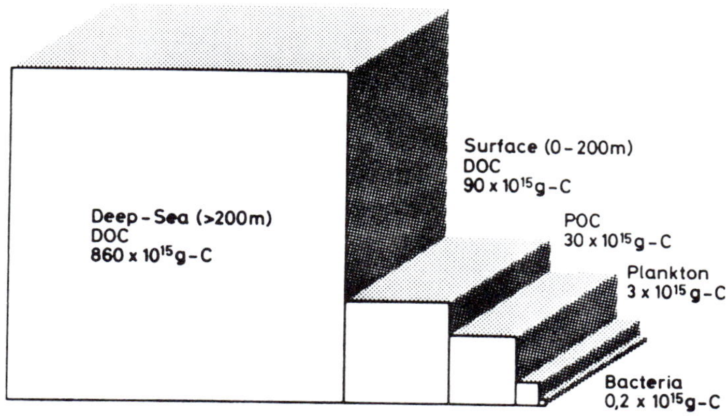

Fig. 2b. Distribution of organic pools in the ocean.

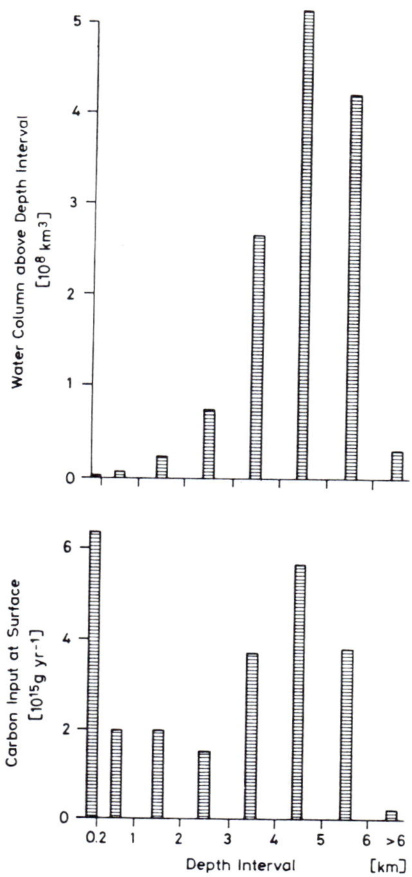

Fig. 3. Distribution of volume of water above different depth intervals in the sea and the inputs of organic carbon to the surface waters above the same depths.

carbon in the sea and by land biomass. In contrast, the oceanic biomass is minimal. Turning to the oceanic carbon pool, the major role is played by the dissolved organic carbon of the deep sea. Although this is a major sink quantitatively, it is rather inertbiologically and appears to have originated mainly from inputs from rivers (Degens and Ittekkot, 1984). Particulate organic carbon, plankton and bacteria contribute to the rest of the oceanic organic carbon pool. Riverine contribution of particulate organic carbon to the open sea is minimal, because it is mainly trapped in estuaries and deltas (Deuser, 1979).

Figure 3 presents data on the distribution of volume of water above different depth intervals, and the inputs of organic carbon to surface waters above the same depths. High inputs of organic matter to a small volume of water over the shelves are registered. This is due to inputs from rivers, not only of particulate carbon, but also of mineral nutrients which increase the primary productivity and organic input in these regions.

Previous estimates of the input of total organic carbon (TOC) to the ocean are in the range 100 - 1000 x 10^{12} gC/a (Duce and Duursma, 1977 ; Richey et al., 1980). The estimates of Richey et al. (1980) are based on measurements made in the Amazon river. A more recent estimate given by Schlesinger and Melack (1981) gives the global TOC transport to be around 400 x 10^{12} g C/a. None of the rivers used in these calculations was studied during all stages of the river flow comprising a hydrological year. A prerequisite for flux calculations is the availability of reliable data on water and sediment discharge.

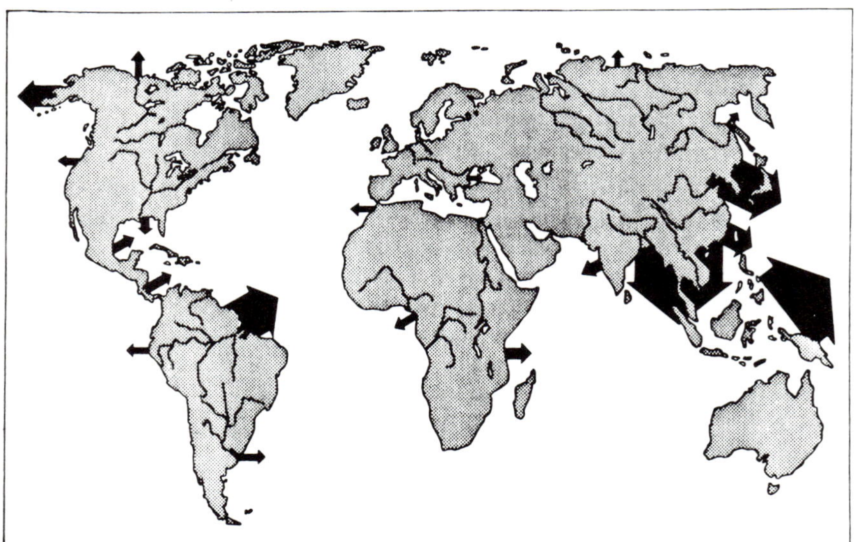

Fig. 4. Annual sediment discharge from various drainage basins of the world (after Milliman & Meade, 1983).

Although adequate information is available from rivers of the temperate regions, they are very scarce for those of tropical rivers. Data reviewed by Milliman and Meade (1983) show that about 70 % of the total sediment discharge to the oceans occur via the Southeast Asian rivers (Fig. 4). Inclusion of sediment data and the organic matter associated with these sediments might change drastically the net POC transport from land to sea.

ORGANIC CARBON TRANSPORT

Data collected over the first two years of our study on DOC and POC were used to calculate annual fluxes of carbon from land to sea. The data are presented not for the individual rivers, but for rivers of a particular continent (Fig. 5). The measurement made in the major rivers of a continent were extrapolated to include the total water and sediment discharge from that particular continent. In terms of quantity maximum POC is being transported by the rivers of Asia, followed by those draining North America, Oceania, Africa and Europe, in that order. South American rivers carry maximum amounts of DOC, followed by the rivers of Asia, North America, the Arctic, Africa, Europe and Oceania.

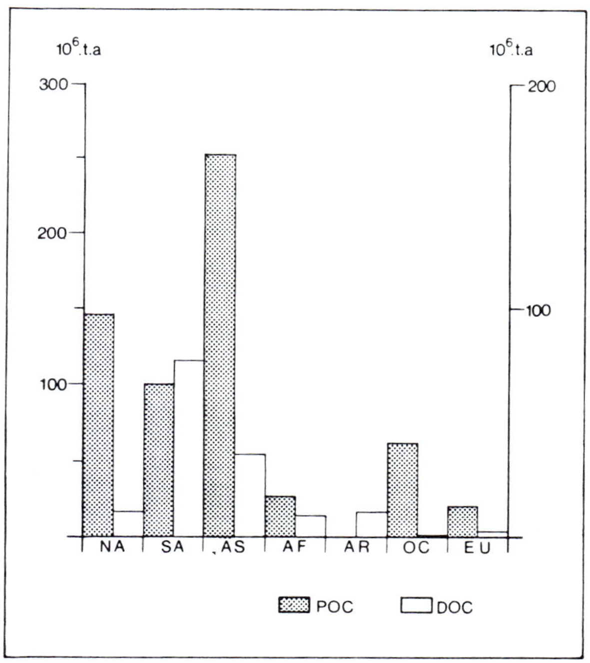

Fig. 5. Annual inputs of organic carbon from the continents via the rivers draining them. NA-North America, SA-South America, AS-Asia, AF-Africa, AR-Arctic USSR, OC-Oceania, EU-Europe.

In terms of POC yield, Asia and Oceania top the list (Fig. 6). Although rivers draining these two regions carry POC in concentrations similar to or less than those measured in the rivers of Arctic and Europe, because of their high sediment yield the POC yield is also the highest. For DOC, maximum yield is for South America and the Arctic, followed by Asia, North America, Africa, Europe and Oceania. High concentrations of DOC in rivers draining the Arctic (above 10 mg/l) contribute to this observed high yield.

The inclusion of the rivers of Asia and Oceania in the flux calculations results in an increase in the quantity of POC carried by the rivers into the ocean compared to DOC.The estimates global transport rates are 0.42 x 10^9t/a for POC and 0.11 x 10^9t/a for DOC (see also Degens et al., 1984).

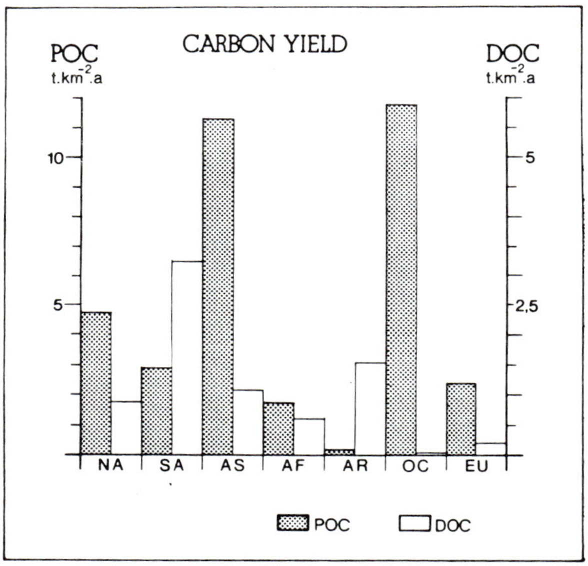

Fig. 6. Annual carbon yield (in tons per square kilometer) for areas depicted in Fig. 3.

SEASONALITY OF ORGANIC CARBON TRANSPORT

The SCOPE/UNEP study has also shed considerable light on seasonal fluctuations in the transport of organic matter. Almost all the rivers studied show such fluctuations (see e.g. Chowdhury et al., 1982 ; Paolini et al., 1982 ; Arain & Khuhawar, 1982). To illustrate the nature of such fluctuations we present data on the Orinocco and the Ganges. In Orinocco the DOC concentrations vary between 5 mg C/l in February and 10 mg C/l in June-July (Fig. 7). The lowest concentrations are observed during September. An increase in suspended load with increasing discharge is also observed. The POC concentrations fluctuate between 0.5 mg C/l and more than 8 mg C/l.Similar fluctuations are also observed in the Ganges (Fig. 8). The concentrations of suspended matter varied between a

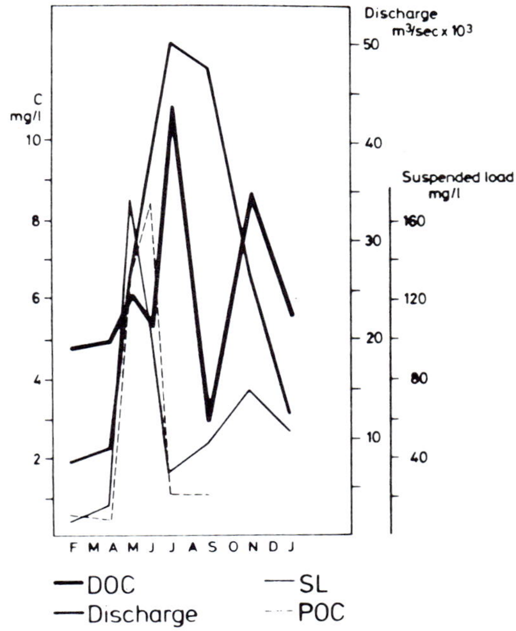

Fig. 7. Seasonal fluctuations in organic matter transport in the Orinocco River.

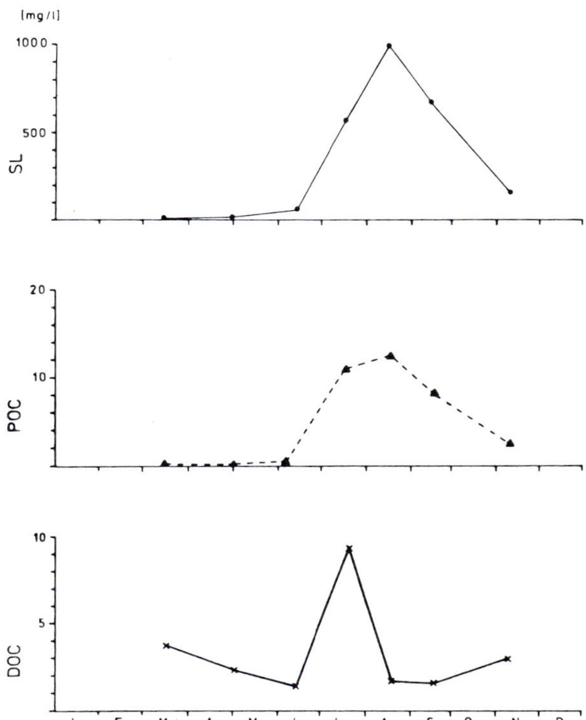

Fig. 8. Seasonal fluctuations DOM transport in the Ganges River.

few milligrams during January-March and 1 mg/l in July. The POC concentrations increased with increasing suspended load and showed variations in the range of a fraction of a milligram to more than 10 mg/l. The maximum DOC concentrations (up to 9 mg C/l) were observed in July-August.

With a view to ascertaining the nature of the transported organic matter, detailed organic geochemical analyses were carried out. The data on sugars and amino acids associated with DOC in the above rivers are given in Figs. 9 and 10. Variations similar to DOC are also observed here. In the Orinocco concentrations of sugars and amino acids varied between 10 and 300 µg/l and 100 and 970 µg/l respectively (Fig. 9). Maximum concentrations were observed in July. The sugar fraction comprises mostly of glucose. Similar trends could also be discerned for the Ganges, with maximum sugar and amino acid concentrations at the rising water stage (Fig. 10).

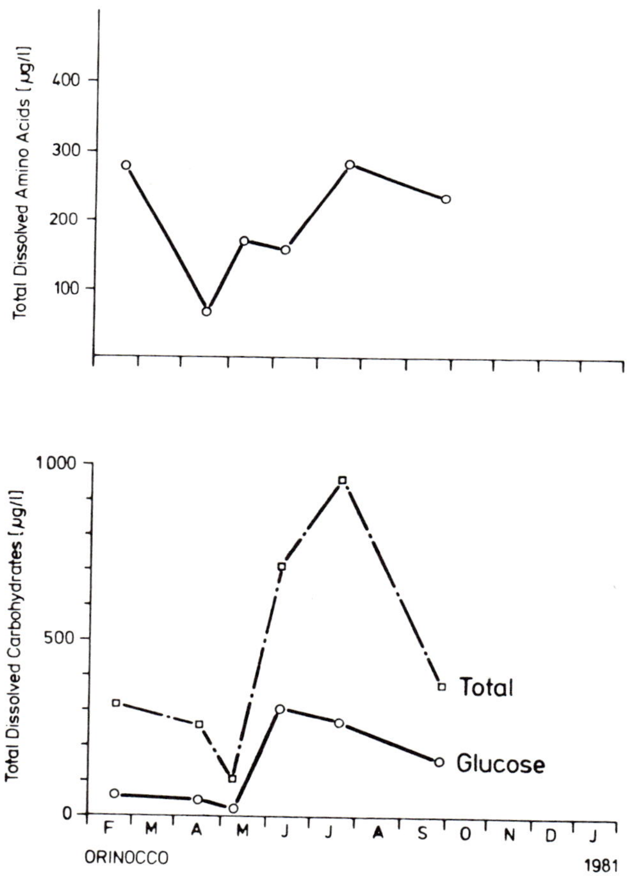

Fig. 9. Seasonal fluctuations in dissolved sugars and amino acids in the Orinocco River.

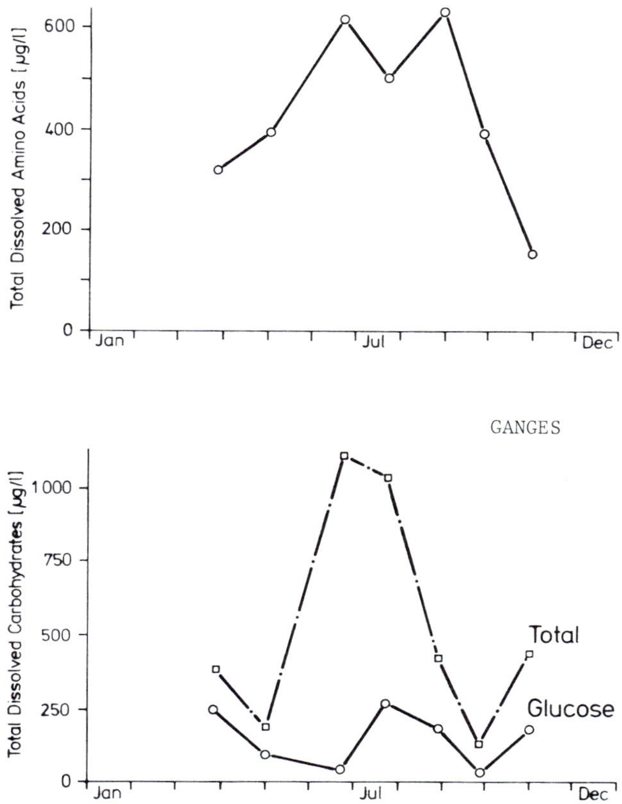

Fig. 10. Seasonal fluctuations in dissolved sugars and amino acids in the Ganges River.

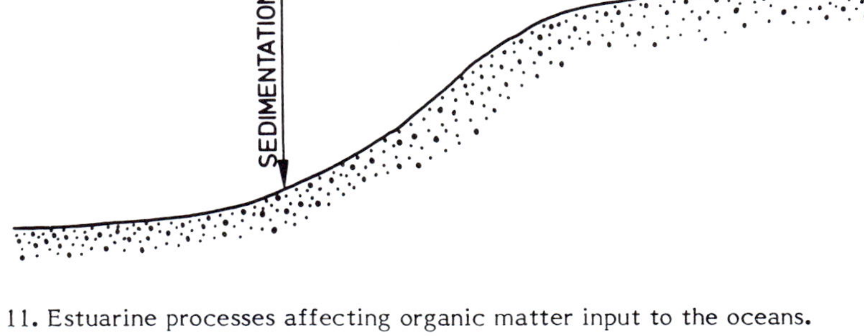

Fig. 11. Estuarine processes affecting organic matter input to the oceans.

The above data indicate that there is a strong seasonality in the transport of materials by the world's rivers (see also Degens et al., 1985). Furthermore, it underscores the importance of collecting samples at least through a hydrological year, especially for flux calculations. During the ongoing SCOPE/UNEP Program determinations of chemical parameters were mainly carried out on samples collected from the rivers in areas above the influence of seawater. However, in calculating the total fluxes or organic matter from river to sea it is imperative to take into account the estuarine processes. These processes, such as flocculation, adsorption, desorption and sedimentation, alter the nature and quantity of organic matter ultimately entering the sea (Fig. 11).
Detailed investigations were carried out in the estuaries of some of the European rivers such as the Elbe, Weser and the Ems (Fig. 12) in order to understand the nature of these processes. The major emphasis during these studies was the behaviour of organic matter during estuarine mixing.

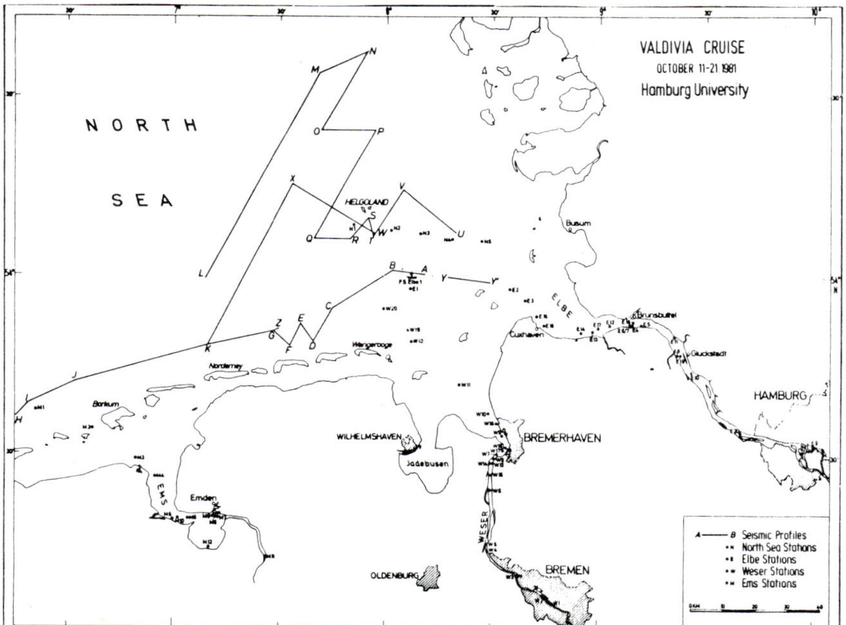

Fig. 12. Map depicting sampling locations in the estuaries of the rivers Elbe, Weser and the Ems Rivers.

The samples were collected from the F/S Valdivia cruise during 1981 (see Degens, 1982). In Figure 13 the distribution of DOC with increasing salinity in the Elbe, Weser and the Ems are depicted. The plots are similar to those reported previously (e.g. Mantoura and Woodward, 1983) and are indicative of conservative behaviour.

The salinity-dependent variations of POC measured in the rivers Elbe and Ems also show decrease with increasing salinity (Fig. 14) (see also Lohse and Michaelis, 1983). However, it is difficult to differentiate between the effect of simple dilution and sedimentation of particles.

Individual constituents of dissolved organic matter such as sugars in the above rivers also show a pattern similar to that observed for dissolved organic matter (Fig. 15). Interestingly for both DOC and dissolved sugars there is a large scatter in concentrations in the low salinity regions (less than 10 °/oo). This is possibly to the complex biogeochemical processes occurring at the contact of river and seawater. Such processes appear to determine the distribution of organic matter in other estuaries too (Eisma et al., 1982, 1983).

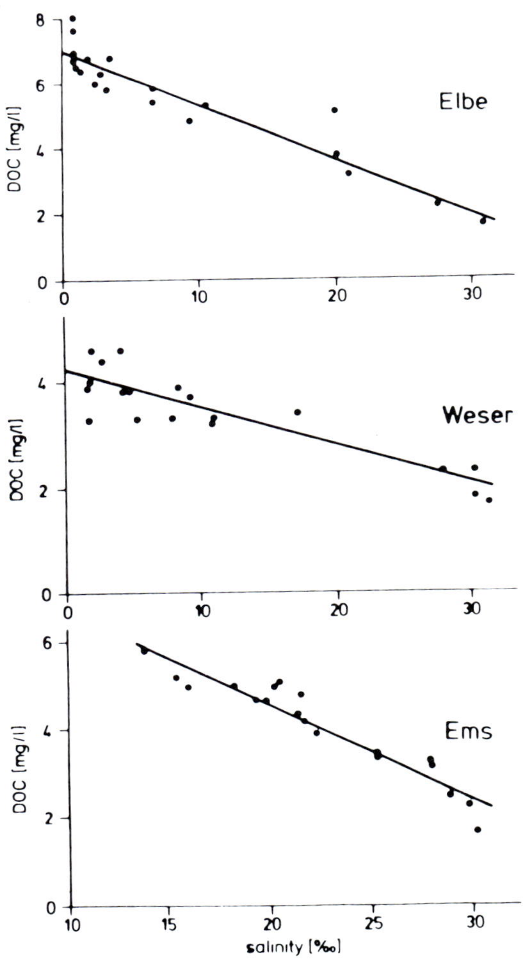

Fig. 13. Plots of salinity vs. DOC in the estuaries of the rivers Elbe, Weser and the Ems.

Estuaries with high inputs of organic waste products may exhibit processes which are different from those described above. Decomposition of organic matter is found to occur in such estuaries. This is seen from studies done on samples collected from the Ems-Dollart estuary (Fig. 12). The samples were collected at intervals of approximately 1°/oo salinity. Dissolved organic carbon and carbohydrates determined on the samples indicate (Fig. 17) that between salinities 0 and 4°/oo there is a net consumption of these two species. This is calculated from the area between the theoretical mixing line and the observed concentrations.

The concentrations of individual sugars (Fig. 17) show that there is a selective degradation of dissolved sugars, with glucose and galactose being mineralized rapidly in the salinity region of 0 to 4°/oo. Beyond this region their distribution is uniform suggesting their stable nature (Laane & Ittekkot, 1983). It appears that the decomposed monomers are in a form which can be easily utilized by heterotrophic organisms.

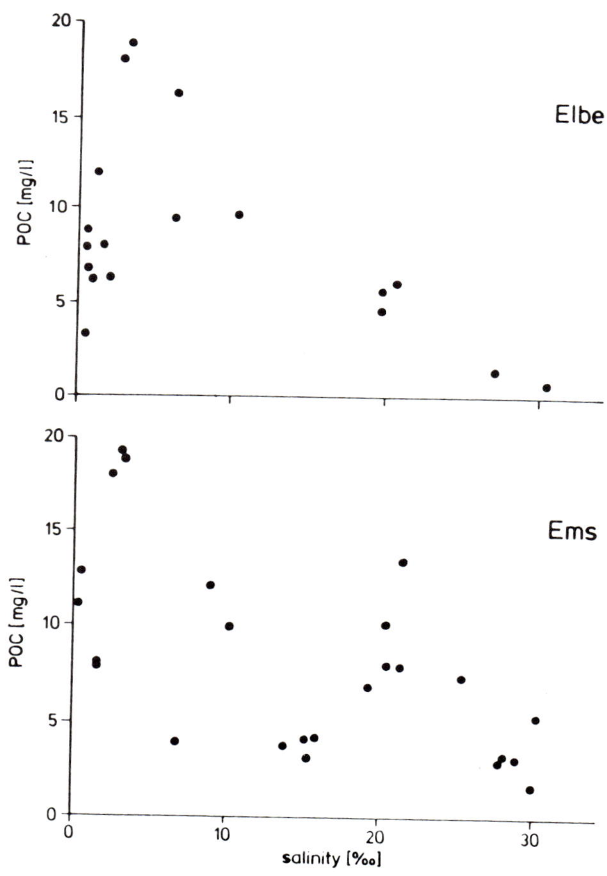

Fig. 14. Plots of salinity vs. POC in the rivers Elbe and the Ems.

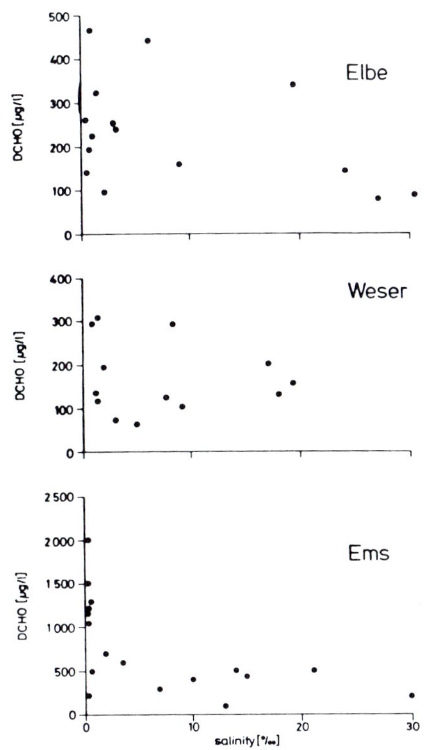

Fig. 15. Plots of salinity vs. dissolved sugars in the rivers Elbe, Weser and Ems.

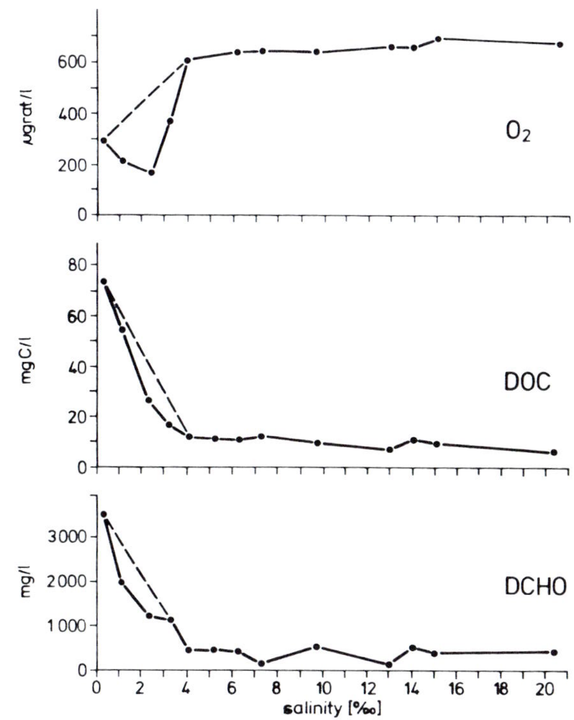

Fig. 16. Plots of salinity vs. DOC, dissolved sugars and oxygen in the Ems-Dollart estuary.

The data presented show that processes controlling the behavior of organic matter in an estuary is dependent, among Figure 16. Plot of salinity vs. DOC, dissolved sugars and oxygenothers, on the nature and quantity of organic input. This will also determine the quantity and nature of organic matter reaching the sea. Rough estimates based on sugars and amino acis in world rivers (Ittekkot et al., 1982) suggest up to 30 % of riverine organic carbon may be labile. Taking these data to be average from world rivers, the amount of DOC entering the oceans is of the order of ca. 0.08 x 10^9 t C/a. Up to 70 % of the POC has a potential for degradation in the coastal marine environment (Degens and Itekkot, 1985).

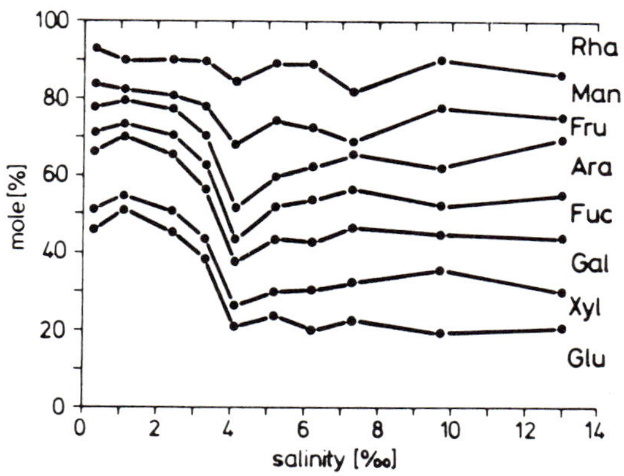

Fig. 17. Plot of salinity vs. individual sugars in the Ems-Dollart.

CONCLUSION

Data collected from an ongoing SCOPE/UNEP program on major world rivers are presented. The total quantity of organic matter carried by the major world rivers are 0.42 x 10^9t C/a for POC and 0.11 x 10^9t C/a for DOC. Significant seasonal fluctuations in both the POC and DOC transport were observed, indicating the importance of sampling through at least one hydrological year. Studies carried out in the estuaries of some North European rivers show DOC to behave conservatively. However, decomposition of organic matter takes place in estuaries with high input of organic matter. On a global scale approximately 30 % of the DOC is labile and may be decomposed.

REFERENCES

Chowdhury, M.I., Safiullah, S., Iqbal Ali, S.M., Mofizuddin, M. and Enamul, S., 1982. Carbon transport in the Ganges and the Brahmaputra : preliminary results. In: E.T. Degens (ed.), Transport of Carbon and Minerals in Major World Rivers. Pt. 1, Mitt. Geol.-Paläont. Inst. Univ. Hamburg, SCOPE/UNEP Sonderbd., pp. 457-468.

Clarke, F.W., 1924. The data of geochemistry. U.S. Geological Survey Bulletin, 770 pp.

Degens, E.T. (ed.), 1982. Transport of Carbon and Minerals in Major World Rivers. Pt. 1. Mitt. Geol.-Paläont. Inst. Univ. Hamburg, SCOPE/UNEP Sonderbd., 52: 766 pp.

Degens, E.T. and Ittekkot, V., 1983. Dissolved organic carbon and overview. In: E.T. Degens et al. (eds), Transport of Carbon and Minerals in Major World Rivers. Pt. 2, Mitt. Geol-Paläont. Inst. Univ. Hamburg, SCOPE/UNEP Sonderbd., 55: 295-314.

Degens, E.T. and Ittekkot, V., 1985. Particulate organic carbon - and overview. In: Degens et al. (eds.), Transport of Carbon and Minerals in Major World Rivers. Pt. 3, Mitt. Geol.-Paläont. Inst. Univ. Hamburg, SCOPE/UNEP Sonderbd. (in press).

Degens, E.T., Kempe, S. and Herrera, R. (eds.), 1985. Transport of Carbon and Minerals in Major World Rivers, Pt. 3. Mitt. Geol.-Paläont. Inst. Univ. Hamburg, SCOPE/UNEP Sonderbd. (in press).

Degens, E.T., Kempe, S. and Ittekkot, V., 1985. Monitoring carbon in world rivers. Environment, 26: 29-33.

Dengens, E.T., Kempe, S. and Soliman, H. (eds), 1983. Transport of Carbon and Minerals in Major World Rivers, Pt. 2. Mitt. Geol.-Paläont. Inst. Univ. Hamburg, SCOPE/UNEP Sonderbd., 55: 535 pp.

Deuser, W.G., 1979. Marine biota, nearshore sediments, and the global carbon balance. Org. Geochem., 1: 243-247.

Duce, R.A. and Duursma, E.K., 1977. Inputs of organic matter to the ocean. Mar. Chem., 5: 319-339.

Eisma, D., Cadee, G.C. and Laane, R., 1982. Supply of suspended matter and particulate and dissolved organic carbon from the Rhine to the coastal North Sea. In: E.T. Degens (ed.), Transport of Carbon and Minerals in Major World Rivers. Pt. 1, Mitt. Geol.-Paläont. Inst. Univ. Hamburg, SCOPE/UNEP Sonderbd., 52: 483-505.

Eisma, D., Boon, J., Gronewegen, R., Ittekkot, V., Kalf, J. and Mook, W.G., 1983. In: E.T. Degens et al. (eds), Transport of Carbon and Minerals in Major World Rivers. Mitt. Geol.-Paläont. Inst. Univ. Hamburg, SCOPE/UNEP, Sonderbd., 55: 295-314.

Garrels, R.M. and McKenzie, G., 1971. Evolution of sedimentary rocks. Chichester, New York, Brisbane, Wiley: 397 pp.

Holeman, J.N., 1968. Sediment yield of major rivers of the world. Water Resources Res., 4: 737-747.

Ittekkot, V., Spitzy, A. and Lammerz, U., 1982. Valdivia Cruise October 1981. Dissolved organic matter in the Elbe, Weser and Ems Rivers and the German Bight. In: E.T. Degens (ed.), Transport of Carbon and Minerals in Major World Rivers. Pt. 1, Mitt. Geol.-Paläont. Inst. Univ. Hamburg, SCOPE/UNEP Sonderbd., 52: 749-764.

Laane, R.W.P.M. and Ittekkot, V., 1983. Behaviour of dissolved organic waste in a part of the Ems-Dollart Estuary : the Dollart. In: E.T. Degens et al (eds), Transport of Carbon and Minerals in Major World Rivers. Pt. 2, Mitt. Geol.-Paläont. Inst. Univ. Hamburg, SCOPE/UNEP Sonderbd., 55: 343-363.

Lohse, J. and Michaelis, W., 1983. Carbohydrates in particulate matter of the Elbe estuary. In: E.T. Degens and S. Kempe (eds), Transport of Carbon and Minerals in Major World Rivers. Pt. 2. Mitt. Geol.-Paläont. Inst. Univ. Hamburg, SCOPE/UNEP Sonderbd., 55: 371-383.

Mantoura, R.F.C. and Woodward, E.M.S., 1983. Conservative behaviour of riverine dissolved organic carbon in the Severn estuary : chemical and geochemical implications. Geochim. Cosmochim. Acta, 47: 1293-1309.

Martin, J.M., Burton, D. and Eisma, D., 1980. River inputs to ocean systems. Switzerland: UNEP and UNESCO.

Meybeck, M., 1982. Carbon, nitrogen and phosphorus transport by major world rivers. Amer. J. Sci., 282: 401-501.

Milliman, J.D. and Meade, R.H., 1983. World-wide delivery of river sediment to the ocean. J. Geol., 91: 1-21.

Paolini, J., Herrera, R. and Nemeth, A., 1983. Hydrochemistry of the Orinocco and Caroni Rivers. In: E.T. Degens and S. Kempe (eds), Transport of Carbon and Minerals in Major World Rivers. Pt. 2, Mitt. Geol.-Paläont. Inst. Univ. Hamburg, SCOPE/UNEP Sonderbd., 55: 223-236.

Richey, J., Brock, J.T., Naiman, R.J. et al., 1980. Organic carbon: oxidation and transport in the Amazon River. Science, 207: 1348-1415.

Schlessinger, W.H. and Melack, J.M., 1981. Transport of organic carbon in the world rivers. Tellus, 31: 172-187.

SURFACE PROPERTIES OF PARTICLES AT THE LAND-SEA BOUNDARY

J.M. MARTIN[1], J.M. MOUCHEL[1] and J. JEDNACAK-BISCAN[2]

[1] Institut de Biogéochimie Marine - UA-386 - E.N.S., 46, rue d'Ulm, 75230 Paris (France)

[2] Laboratory of Electrochemistry, Rudjer Boskovic Institute, POB 1013, Zabreb (Yugoslavia)

ABSTRACT

Martin, J.M., Mouchel, J.M. and Jednacak-Biscan, J., 1986. Surface properties of particles at the land-sea boundary. In: P. Lasserre and J.M. Martin (eds), Biogeochemical Processes at the Land-Sea Boundary. Elsevier, Amsterdam.

The determination of surface properties of particles is an important key to understanding interactions of trace elements and organic compounds between particulate and dissolved phases in estuarine and coastal systems. Specific surface area (SSA), cationic exchange capacity (CEC) and heat of immersion (ΔH) have been measured on native and treated suspended sediment and after oxidation with 15% H_2O_2. SSA and ΔH have also been measured on samples leached with NaOH and Na-dithionite in order to remove amorphous aluminosilicates.

Results indicated that SSA is controlled by amorphous oxides and is related to the Al content of samples and their granulometry, suggesting an almost uniform coating of the detrital core by amorphous oxides. On the other hand, CEC and ΔH, which appear to be highly correlated, are controlled by both organic matter and amorphous oxides.

Studies on electrophoretic mobility have provided additional data on the excess of charge at the interface between suspended matter and electrolytic medium. Particles in suspension in fresh, sea and estuarie waters appear ubiquitously to exhibit a small range of negative surface charge. This uniformity is attributed to the presence of organic surface coatings on the particles.

INTRODUCTION

An estuary provides a particularly varied chemical environment which is characterized by strong ionic strength and pH gradients ranging from river to sea end members values. In addition one can usually observe a slight pH and dissolved oxygen minimum in the estuary itself which is enhanced in polluted estuaries.

With regards to solid phase, most of the estuaries are characterized by a "turbidity maximum". The fresh water flow generates an entrainment from the bottom water, the sea water which is entrained from the lower layer by this flow being compensated by a residual landward flow along the bottom. The river-borne particles which settle in this bottom layer are taken up by the residual landward current and trapped in the middle part of the estuary. This process greatly enhances the residence time of particles, especially in tidal estuaries.

Moreover, both river induced up-welling and river discharge of nutrients create a fertile environment which enhances the primary production of organic matter in off-shore direction of estuaries. It can be predicted from these observations that heterogeneous reactions between dissolved and both mineral phase and biota will be predominant in estuaries and coastal zones. These reactions will primarily affect those elements and compounds which are located at the particulate surface. The determination of surface properties of particles appear to be an important key to understand the interactions of trace elements and organic compounds between particulate and dissolved phases in estuarine and coastal systems.

Specific surface area (SSA), cationic exchange capacity (CEC) and Heat of immersion (ΔH) have been measured on native untreated suspended sediment and after oxidation with 15% aqueous solution of H_2O_2. In addition, SSA and Heat of immersion have been measured on samples which have been leached with NaOH and Na-dithionite in order to remove amorphous aluminosilicates.

MATERIALS AND METHODS

The estuaries chosen for the studies were the Gironde and Loire estuaries (France) both caracterized by relatively long flushing and residence times (Martin, Mouchel and Thomas, 1984) as well as by a well developped turbidity maximum. Their general characteristics have been described previously (Allen, 1972; Barbaroux, 1980; Elbaz-Poulichet et al., 1982, 1984).

Samples were taken during 3 surveys in the Gironde, selected during different hydrological conditions and 1 survey in the Loire. Large volume samples have been either continuously centrifugated (11 000 or 32 000 rpm) or tangentially filtrated in order to recover the suspended matter for radionuclide and surface properties analysis. These particulate samples have been lyophilized and crushed in an agathe mortar.

A selective leaching procedure, adapted from Hashimoto and Jackson (1960) and Jackson and Mehra (1960) was used to evaluate Fe, Mn, Al and Si associated with the "amorphous" aluminosilicates. It is based on two time-controlled hot digestions by 0.5 N NaOH, followed by a reduction step using the dithionite-citrate system buffered in $NaHCO_3$. This procedure was supposed to extract poorly cristallized oxides and hydrous oxides without any significant destruction of the well cristallized pure minerals (Elbaz et al., 1982). The organic coating was removed after treatment with 15 % H_2O_2 at 150°C for 12 hours. P.O.C. was determined using a LECO carbon analyser.

Specific surface areas were determined by the volumetric method of adsorption of argon at liquid N_2 temperature, using the BET equation (Parfitt and Sing, 1976).

Heats of immersion were measured using a batch calorimeter (Pravdic, 1976). It was an isoperibolic instrument with a thermistor detector of an average sensitivity of $\pm$ 1 mJ (which corresponds to an estimated ± 3 $\mu J/cm^2$). A sediment sample sealed in a small glass bulb was brought into the microcalorimeter cell and carefully equilibrated to a temperature of 25 ± 0.2°C (kept, however stable to + 0.005°C). The bulb was then crushed and the heat of immersion automatically recorded. In each experiment, the heat evolved was calibrated by means of an input of a measured amount of current to a small in situ heater (Juracic and Pravdic, 1981).

The electrophoretic mobility of natural suspended sediments has been measured on the field a few hours after sampling using a Pen Kem's Model 501 Laser Zee Meter which uses a rotating prism design enabling simultaneous measurements of many particles.

Model particle mobility has been determinated with the Tiselius method (Tiselius, 1937, 1938). This method also allows the integration of the mobility of a large number of particles even if the refractive index is very close to that of the electrolyte medium, allowing to minimize the experimental errors inherent to the classical microelectrophoretic techniques. The electrophoretic mobilities will not be transformed into surface charges because the theoretical relationship between these parameters is highly dependant on the particle radius of curvature and the electrolyte concentration in the vicinity of the particle (Hunter and Wright, 1971). For both methods, the analytical error falls below 5 %, however, it increases up to 10 % for natural composite samples and/or low mobilities.

C.E.C. was determined by using NH_4^+, saturation method (Busenberg and Clemency, 1973) and specific ammonia electrode (Orion Research Inc., Cambridge). The sediment samples (ca. 150 mg) were first saturated in 1N ammonium acetate solution overnight, then filtered through 0.45 um filters and leached with 1N NH_4Cl. Excess ammonium salts were removed by washing with isopropyl alcohol. A final 80°C drying made the samples ready for C.E.C. determination. Available ammonium was displaced in 0.1 M NaOH, the corresponding potential was measured. Calibration measurements were performed with NH_4Cl and NaOH. The reproductibility of the electrode itself is ± 2 %, and the precision obtained on replicate samples was up to ± 5 %.

RESULTS AND DISCUSSION

Electrophoretic mobility

As pointed out by Hunter (1981) many of the important properties of colloidal systems are determined directly or indirectly by the electrical charge (or potential) on the particles. Adsorption of ions and dipolar molecules is determined by, and also determines, this charge and potential distribution. The potential distribution itself determines the energy of interactions between the particles, and is in many cases responsible for the stability of particles towards coagulation and for many aspects of the flow stability of the colloidal suspension. Controversial potential data have been obtained during the last decade (Pravdic, 1970 ; Martin, Jednacak and Pravdic, 1971 ; Hunter and Liss, 1982) but it is now established that river suspended matter is negatively charged with a slight decrease of the electrophoretic mobility during estuarine mixing. We have repeated these experiments in the two studied estuaries (Fig. 1). The electrophoretic mobilities appear almost similar in the two estuaries despite different mineralogical compositions. Although the electrophoretic mobility of many minerals is negative at natural pH, this suggests, as pointed out by Hunter and Liss (1982), that some coatings control the surface charge of the particles. The electrophoretic mobilities in the Gironde and the Loire estuaries indicate an uniformly negative charge from the river to the ocean. The mobility remains constant over the whole tidal estuary located above the salt intrusion. This suggests that those particles which have been transported landwards from the middle estuary to this area are quickly reequilibrated with the surrounding water. The electrophoretic mobility decreases sharply by a factor to two during the first step of mixing. Then, it gradually decreases for higher chlorinities, but remains negative as previously mentioned. These data fall in the range of those reported for other estuaries (Hunter and Liss, 1982 ; Loder and Liss, 1982).

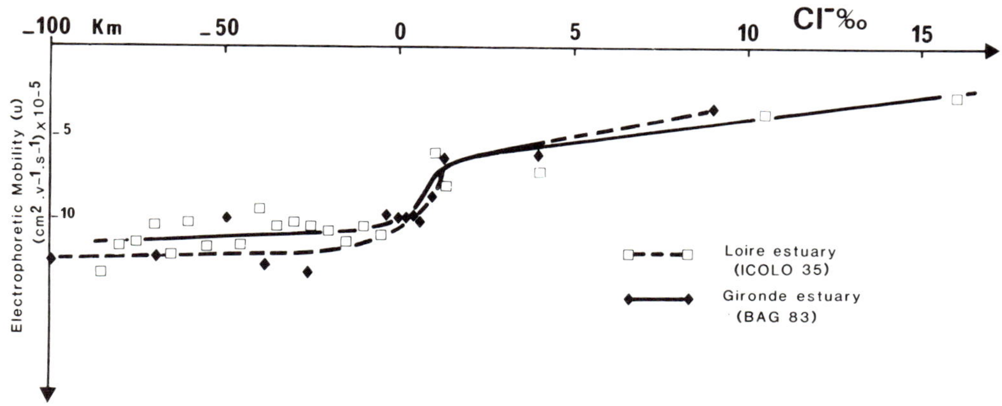

Fig. 1. Evolution of the electrophoretic mobilities of suspended sediment from the Gironde and Loire estuaries (France). May 1983.

We have tried to specify the influence of different physico-chemical parameters upon the electrophoretic mobility, using model particles such as silica (Aerosil 380 - Degussa) which is characterized by a charge very close to that of natural suspended sediment from Loire and Gironde.

a) Influence of pH. Figure 2 shows the evolution of the electrophoretic mobility (u_E) with the pH variation in a 10^{-2} mole/l NaCl solution. In such a medium, the isoelectric point (IEP) has been observed at a pH = 3.4 ± 0.5. This value is in good agreement with other results (Bolt, 1957 ; Ahmed, 1966). An IEP, due to the acidic properties of the surfaces, is commonly observed for other oxides, however very few data are available for natural suspensions. Hunter (1979) determined the influence of pH on the electrophoretic mobility of synthetic particles coated with sea water organics, he suggested the major influence of carboxylic and phenolic groups. It is clear from these data that the slight pH decrease usually observed along the Gironde or Loire estuaries cannot explain the charge decrease observed in the low chlorinity zone but can contribute to this variation.

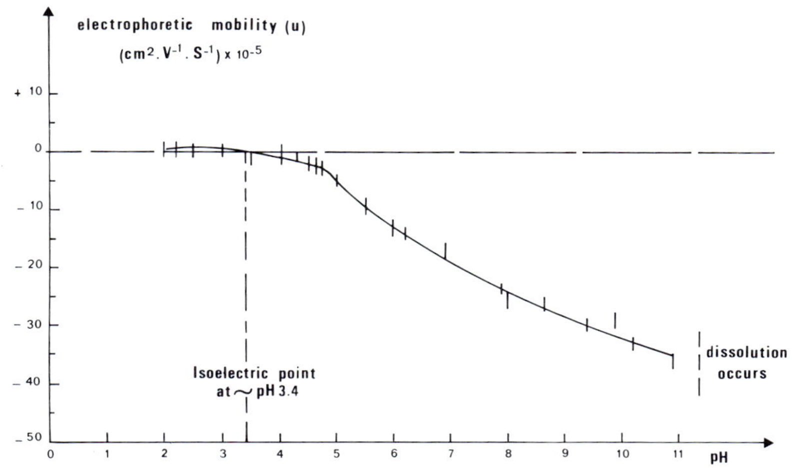

Fig. 2. Electrophretic mobility with the pH variation in a NaCl solution (10^{-2} mole/l)

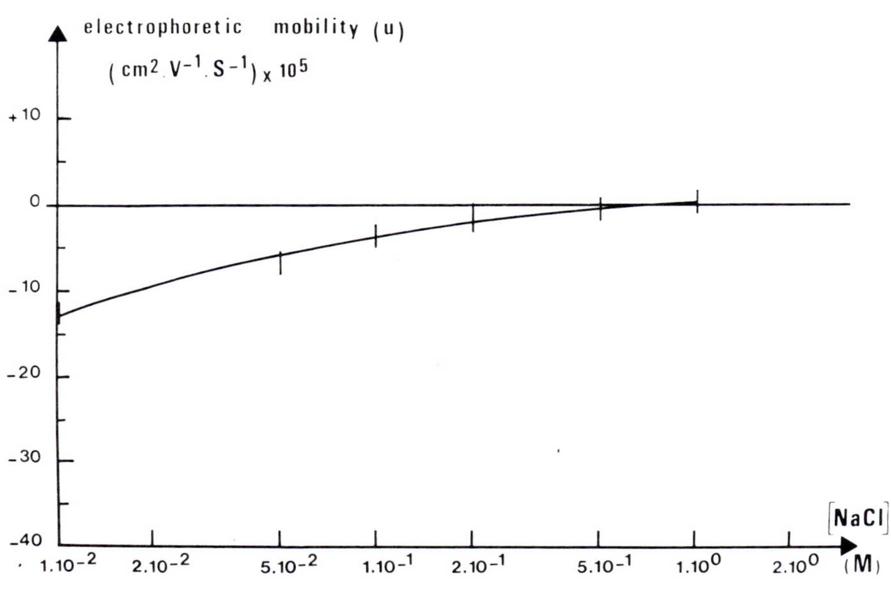

Fig. 3. Electrophoretic mobility of Aerosil-380 silica particles as a function of NaCl concentration (pH = 6).

b) Influence of NaCl concentration. Figure 3 shows the evolution of the electrophoretic mobility of the silica as a function of the NaCl concentration in solution at pH = 6. For such pH, the silica sample is strongly dissociated, and the effect of the negative charge of the particles can only be neutralized at high NaCl concentrations, between 0.5 and 1 M/l. Data are very similar to these obtained for suspended matter in estuaries.

c) Influence of organic substances. The importance of adsorbed organic matter in affecting the electrical charge of particles in sea water has been demonstrated by Martin, Jednacak and Pravdic (1971), Neihof and Loeb (1972, 1974). Recent work by Hunter and Liss (1979, 1982) has confirmed this importance. Finally Loder and Liss (1982, 1985) and Hunter (1980) clearly demonstrate that, whenever iron oxides or alumina are considered, their specific positive charges are changed radically to usual negative values by the formation of organic coatings in sea or estuarine waters (Fig. 4). They also pointed out the specific adsorption of major cations on these positively charged particles, since their electrophoretic mobility increases with the salinity when the sea water is first UV-oxidized, whilst a Gouy-Chapman model would have predicted a decrease.

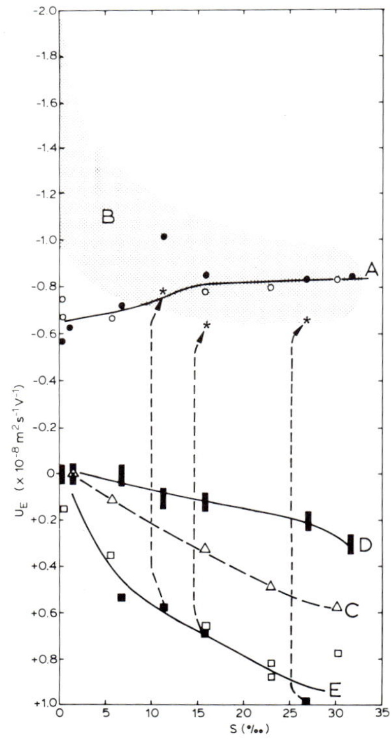

Fig. 4. Electrophoretic mobilities (U_E) of natural (untreated) - curve A - and treated particles as a function of salinity (S°/oo) for two sets of samples from Keithing Burn (KB 1 : open symbols - 31 March 1982 ; KB 2 : closed symbols - 30 June 1982). Shaded area B indicates the spread of results from other estuaries (redrawn from Fig. 3 of Hunter and Liss 1979). Curve D - suspended particles centrifuged and resuspended in UV- oxidized sample supernatant and then UV-oxidized. Curve C - natural samples (particles plus supernatant) UV-oxidized. Curve E - sample supernatant UV-oxidized to form new particles (UV-PPT). Several UV-PPT samples from KB2 were centrifuged and the particles resuspended in their original untreated sample supernatant. The resulting changes in U_E are indicated by the dashed lines (asterisks - final values). Keithing Burn suspended matter is mostly composed of iron oxides (after Loder and Liss, 1985).

Tipping (1981) showed that the adsorption of terrestrial humic substances could reverse the positive electrophoretic mobility of iron oxides. Thus, both terrestrial and marine organics are strong adsorbates, able to reverse zeta potential.

d) Application to estuarine data. Data for the Gironde and Loire estuaries are reported in Fig. 5, with data from Hunter and Liss (1982) for the Conwy estuary. Clearly the main factor explaining the evolution of zeta potential is salinity, while the globally negative charge is due to both the negative charge of most structural constituents of estuarine particulate matter and to organic coatings.

These data are difficult to relate to an evolution of electrostatic term of ion adsorption. It is commonly admitted that, for colloidal systems, the zeta potential obtained from the electrophoretic mobility and the Smoluchkowski equation, possibly corrected by a size or shape factor, is a good approximation for the OHP potential at relatively high ionic strength (> 10^{-3} M) (Foissy et al., 1982). Recently, a model based on this concept has been successfully used to interpret adsorption date as a function of pH for anions onto iron oxide (Haussmann and Anderson, 1985). In the case of pH, at constant ionic strength, the great variability of the electrokinetic potential has to be related to a variation of the OHP potential. However, in the case of salinity effect, and for natural suspended matter the slight evolution which has been observed in several estuaries is not

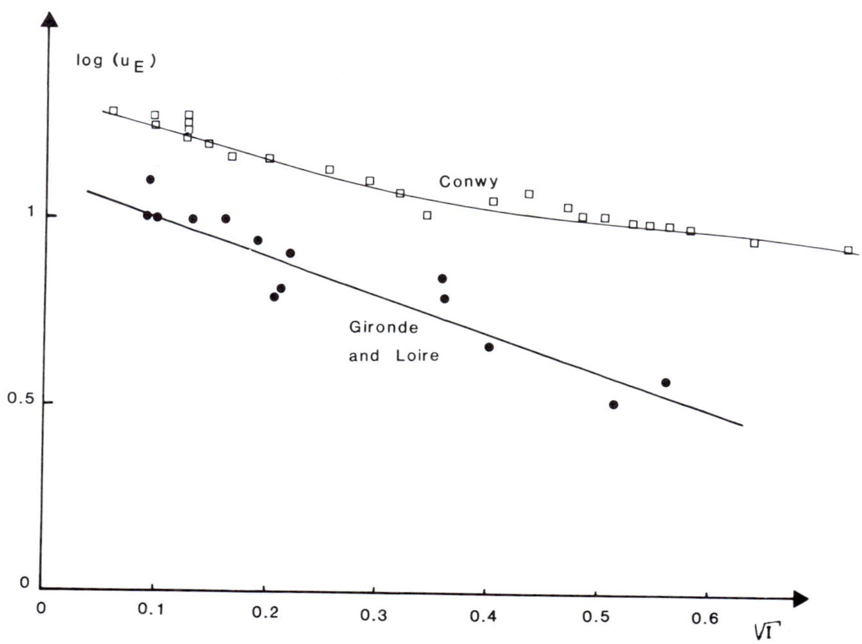

Fig. 5. Plot of the logarithm of the electrophoretic mobility of natural particles in the Gironde, Loire and Conwy estuaries (Date for Conwy from Hunter and Liss, 1982).

necessarily due to a variation of the OHP potential because of the numerous incertitudes related to the Smoluchkowski equation and because a very small distance between the OHP and the plane of shear (2 Å to 7 Å) in a simple Gouy-Chapman model could the explain the evolution. A careful study by Smith (1976) demonstrated that there was no evidence of the shear plane being at a finite distance from the OHP to a sensitivity of order 5 Å. Thus, it is not clear whether the evolution of the electrophoretic mobility is related to an evolution of the electrostatic term of the interactions between dissolved species and particulate matter.

In situ estuarine measurements of electrophoretic mobility would be interestingly completed by estimations of the effect of ionic strength and of some specified dissolved components of sea water on the sampled particles.

2. Specific surface area

The results of two surveys are reported on Fig. 6, S.S.A. appears to be lower in the river part, especially for the Gironde estuary. It usually increases as soon as the tidal influence is observed with a clear maximum located around the salt intrusion. Many processes can affect particulate matter in this part of estuaries. The increase could be due to the destruction of some flocs formed upstream as observed by Eisma (personal communication) or to a chemical evolution of the superficial composition of particles as discussed below.

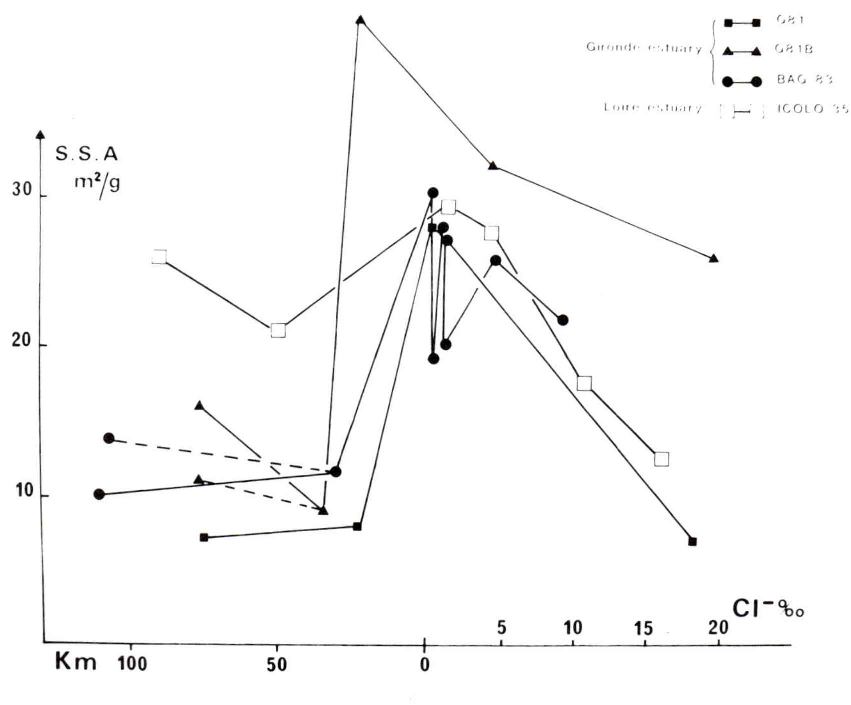

Fig. 6. Longitudinal evolution of S.S.A., in the Gironde and Loire estuaries.

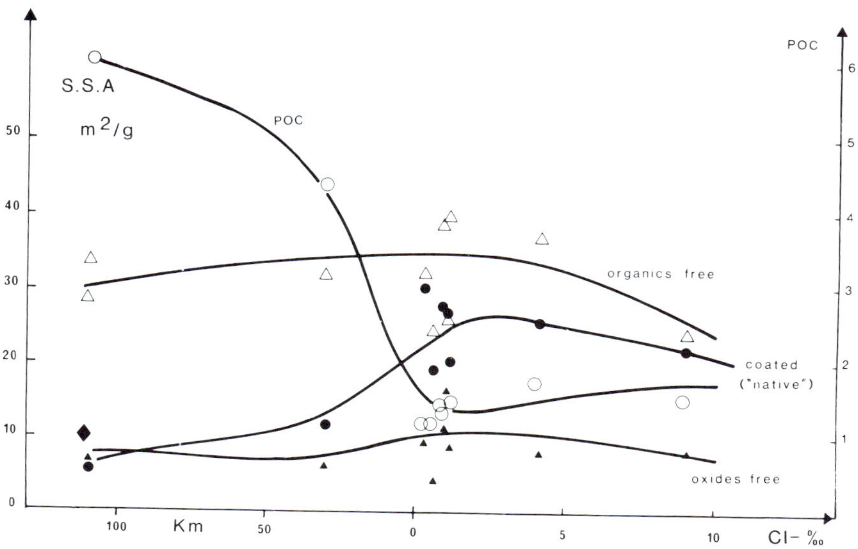

7a: Gironde estuary.

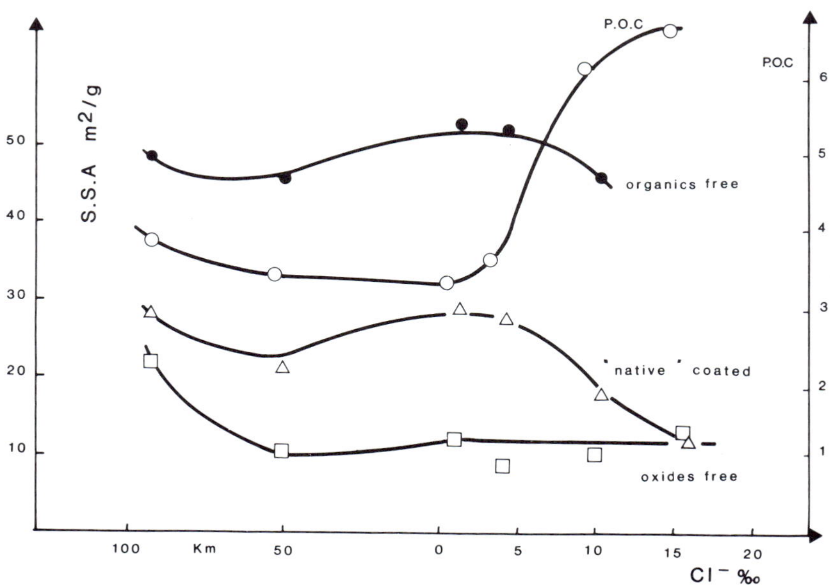

7b : Loire estuary.

Fig. 7. Specific surface area (BET Argon adsorption) of native, organics free and amorphous oxides free samples.

Fig. 8 shows the evolution of the maximum floc size in the Gironde estuary, an important decrease appears around the salt intrusion (R. Gibbs personal communication). The maximum used to be associated with an increase of turbidity so that the S.S.A. per liter is even more pronounced. This maximum could greatly enhance the availability of adsorption sites for dissolved trace metals and radionuclides. The maximum observed in the upper estuarỳ is followed by a decrease in the lower part of both Loire and Gironde. Possible segregations of chemical and mineralogical composition of suspended sediments could partly explain this evolution. Systematic analysis of clay mineralogy as well as quartz and carbonate content did not reveal any significant trend along the estuary.

The aluminium content is commonly used as an indicator of the granulometry of the sample, since most of it is involved in the alumino-silicates which are well known to be the most important part of the smaller fraction. A rather good correlation (Fig. 9) can be found between S.S.A. and Al content.
This confirms the utility of normalizing to Al to compare trace metals or radionuclides concentrations even if one assumes that they are located at the particle surface. Notice also that the normalization can introduce some bias in comparing estuarine and riverine data (in the case of the Gironde), since all the riverine dots are gathered (circle) under the main correlation line.

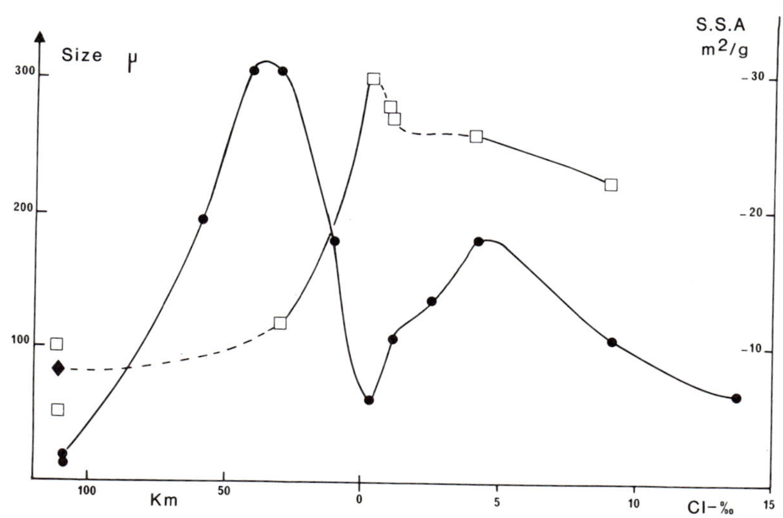

Fig. 8. Comparative evolution of the maximum floc size and the specific surface area in the Gironde estuary.

In order to investigate if surface coatings such as hydrous oxides and/or organic matter were involved in the actual S.S.A. distribution, the analysis have been repeated with organics free and amorphous free samples.

The S.S.A. raises to higher values if the organic matter is stripped of the samples. The data issued from the Gironde estuary (Fig. 7a) demonstrate that the P.O.C. is directly related to an increase of S.S.A.. This means that the organic matter somehow blocks the available sites for Árgon molecules. In the Loire (Fig. 7b), the phenomenon is less observable ; the increase of P.O.C. at sea end member is not correlated with a greaterincrease of the S.S.A. after treatment. But it has to be noticed that organic matter composition of this sample, is quite different from river organic matter as shown by the ^{13}C data (R. Letolle, personal communication) this difference can explain its lower activity in blocking the available sites of the inorganic matrix.

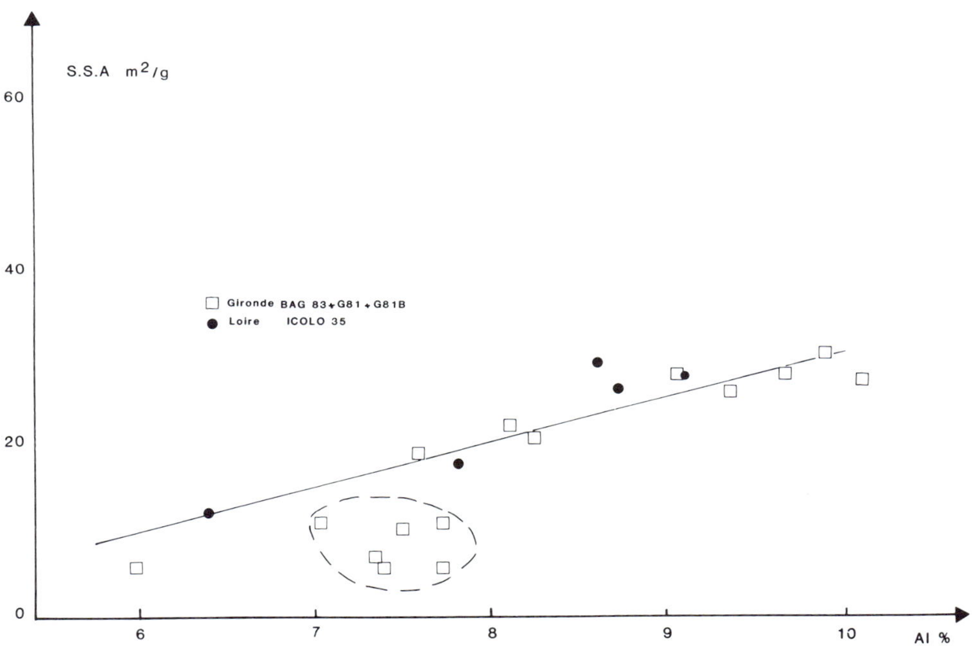

Fig. 9. Correlation between specific surface area and Al content of the samples. Upstream Gironde data are circled.

Marine sediments as well as deposited estuarine sediments never show such differences. Usually, the organic matter content is lower in estuarine than in river sediment, and S.S.A. is always the same or slighthly lower from treated as compared to untreated deposited and marine sediments. This supports the idea that the organic coating of the river particulate matter drastically changes its character depending upon the environment, modifying its adsorptive surface properties for trace metals and radionuclides. The generally negative influence of P.O.C. on the S.S.A. has been demonstrated ; it should be interesting now to relate the organics free S.S.A. to some other parameter characterizing the inorganic core. The amorphous free samples, which are probably also partly organic free, show a considerably lower S.S.A. (Fig. 7), demonstrating that, in as much the leaching is specific, the amorphous oxides greatly contribute to the S.S.A.. The amorphous free samples show an almost constant S.S.A. over the estuary (= $10 m^2/g$). It is higher than the cristallized SiO_2 S.S.A. ($3.3\ m^2/g$, Benjamin, 1980) but lower than the illite S.S.A. ($70\ m^2/g$, O'Connor, 1975). Globally, the "amorphous specific" treatment removes almost 1.2 % Fe_2O_3, 0.8 % Al_2O_3, 4-8 % SiO_2 (expressed as percentages of the total mass of the sample). These percentages are not correlated to the global aluminium or silicum content of the sample showing at least some specificity of the leaching.

"Organics free" S.S.A. should be due to what is removed by the leaching. Indeed, there is a global positive correlation between the leached quantities of "amorphous" oxides and the organics free S.S.A., especially in the case of iron oxide (Fig. 10). Existing data of the litterature on amorphous oxides S.S.A. (Davis, 1978) cannot account for the gap between organics free and amorphous free samples, but the S.S.A. of a pure phase can hardly be compared to the S.S.A. of the same phase coated around an inert core.

It can be concluded that the S.S.A. of estuarine suspended matter is mainly due to amorphous oxides coatings, among them, the iron oxides could be predominant, and the organic coatings behave as a mask for Argon molecules during BET adsorption. A model particule can be considered as an inorganic detritic core surrounded by an active amorphous layer relatively to gas adsorption, partly recovered by a less active organic film.

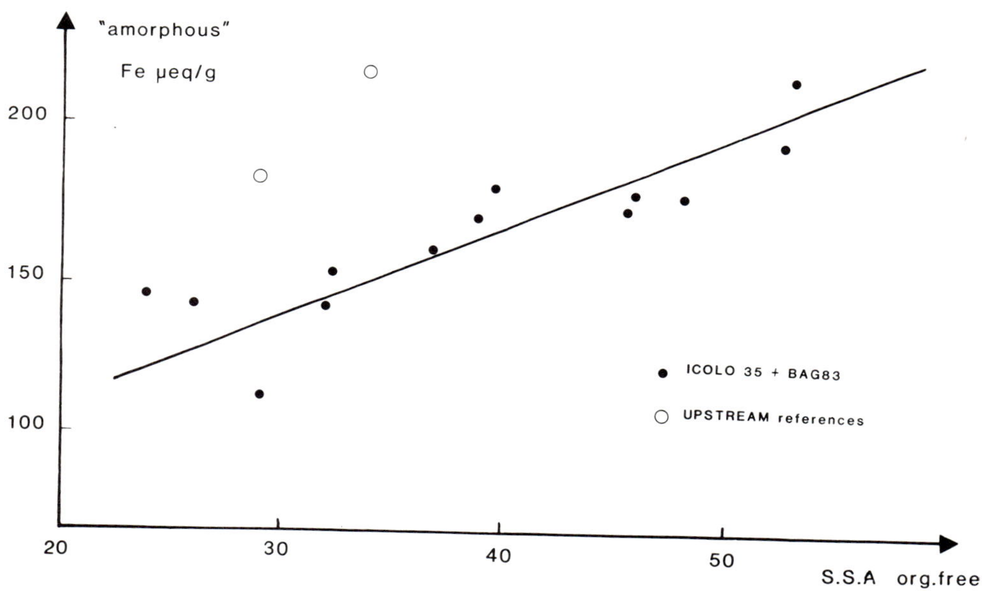

Fig. 10. Correlation between the organics free specific surface area and the operationally defined "amorphous" iron content of the samples.

3. Cation Exchange Capacity

Cation exchange capacity provides an additional parameter to understand the behaviour of some trace metals and radionuclides in estuaries. Keeping in mind the method of determination (NH_4 saturation and release by a strong base), it provides an estimation of the number of sites available for ion exchange or for some other process involved with the same surface active sites. It is now clear that such active sites exist in both the mineral phase of the sediment and in the organic phase. This later contributes to the total CEC from 25 to 80 % (Schnitzer, 1965 ; Rashid, 1969 ; Hunt, 1981). This means that the organic coatings may have a specific CEC. Our results support such findings.

The values obtained in Gironde and Loire are shown Fig. 11. The Loire appears slightly higher than Gironde but both sets of data correspond to slightly higher values than theses measured by Hunt (1981) for Long Island Sound at the sediment surface . In general, there are no systematic changes from the river to the sea, except a possible maximum in the estuary as compared to the two end-members. Whenever the values are normalized to the surface areas, there is no more differences between Loire and Gironde and all the curves indicate a minimum within the estuary.

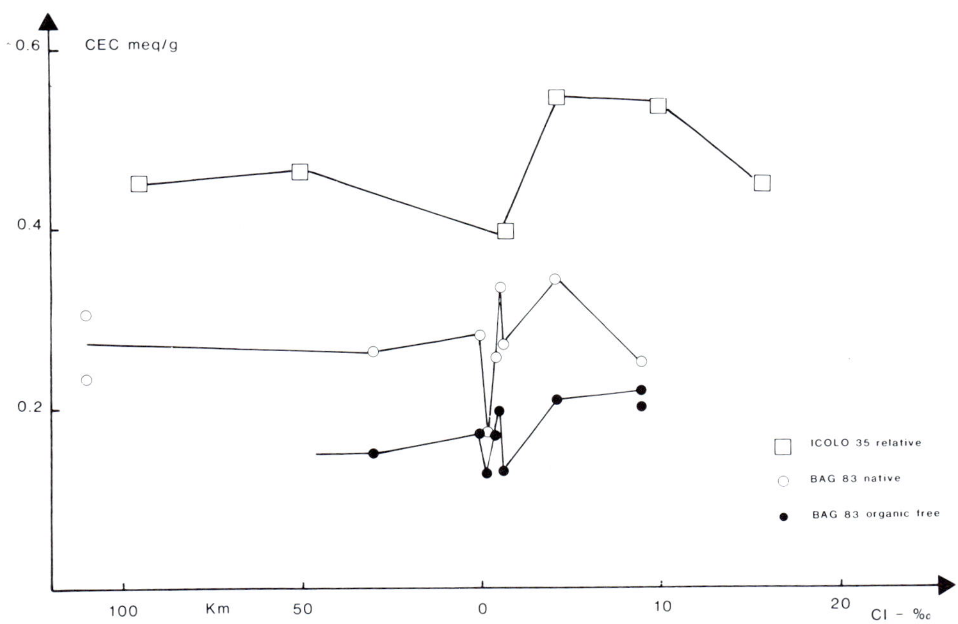

Fig. 11. Cation exchange capacity (NH_4 saturation method) in the Gironde and Loire estuaries.

As in the case of S.S.A., we measured the difference between the "native" and "organics free" samples. It has to be reminded that the measurements were performed with dry samples (80°C), so that a certain fraction of organic sites and amorphous sites may disappear. However, the global effect of the organic part on CEC is clearly expressed. Fig. 10 shows that the CEC values for organics free samples are systematically lower than those of native samples. The residual values for CEC after the removal of the organics (15-20 meq/100 g) should correspond to the CEC of the mineral phase. However, it has to be noticed that the warm oxidation attacks the subsurface layer mainly consisting of amorphous oxides. Therefore, it can not be a relevant indication of mineral phase contribution to the total CEC.

4. Heat of immersion

The heat that is evolved on immersing a solid in a liquid yields a picture of its surface including all its heterogeneities. In many geochemical problems involving solid-liquid interactions, the heat of immersion technique provides a very promising method. The heat of immersion is the enthalpy change ΔH observed when a solid surface is suddenly immersed into some liquid and can be expressed as an energy interaction by surface area, describing the energetics of adsorption of a solute with the particulate surface (Zettelmoyer, 1965 ; Pravdic, Jednacak-Biscan and Juracic, 1981).

Whenever the liquid used is water, ΔH will give some idea of the hydrophilicity of the

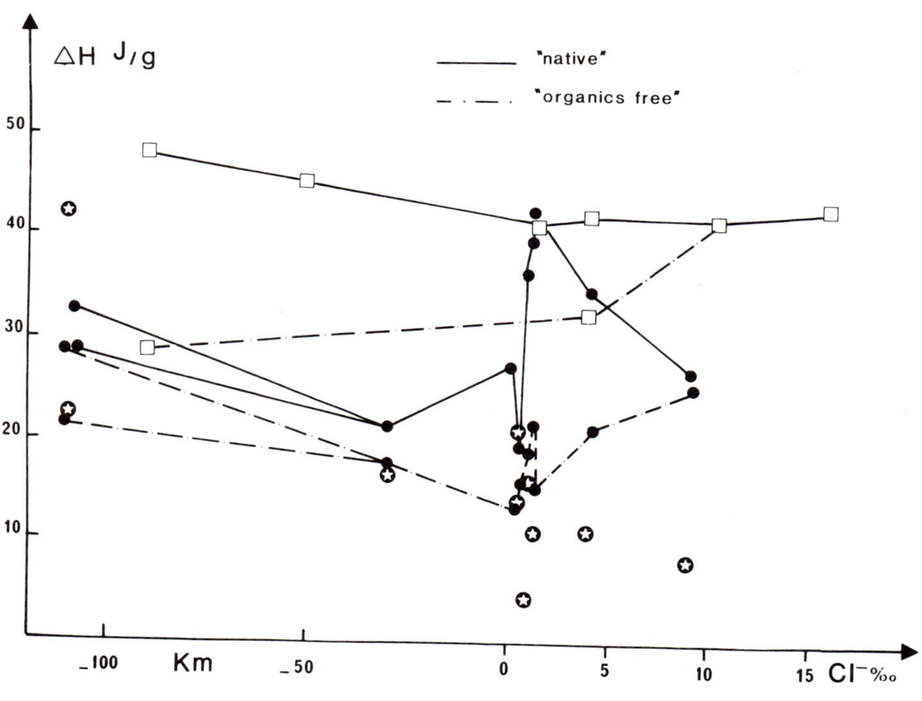

Fig. 12. Evolution of the heat of immersion in the Gironde (closed circles) and Loire (open squares) estuaries. Stars : "amorphorus free " samples, Gironde estuary.

solid surface and its resulting affinity for a large variety of organic pollutants and natural ligands. The higher the surface hydrophilicity the higher is the energy of wetting per unit surface.

The general picture of data for the two estuaries is shown in figure 12. It can be seen that the heat of immersion data follow the CEC data. A highly significant correlation exists between these two sets of data, including native and organics free samples (Fig. 13). The slope of the straight line can give us some information about the kind of processes which ammonium adsorption. In this case the energy involved is 105 kJ/mole H_2O. The energies of surface hydration of oxides are 80-105 kJ/mole for TiO_2, 27 kJ/mole for SiO_2 (Morimoto, 1968) or 84 kJ/mole for Al_2O_3 (Fubini et al., 1978). The heats of wetting as found in our results are higher, suggesting more hydrophilic interactions. Numerous processes can take place when wetting clay minerals with water, such as swelling, ion exchange, chemisorption or chemical interactions (Zettlemoyer, 1965), but they can hardly be resolved in the case of natural inhomogenious sediments. Globally, the value of 105 kJ/mole corresponds to strong physical sorption or even chemisorption.

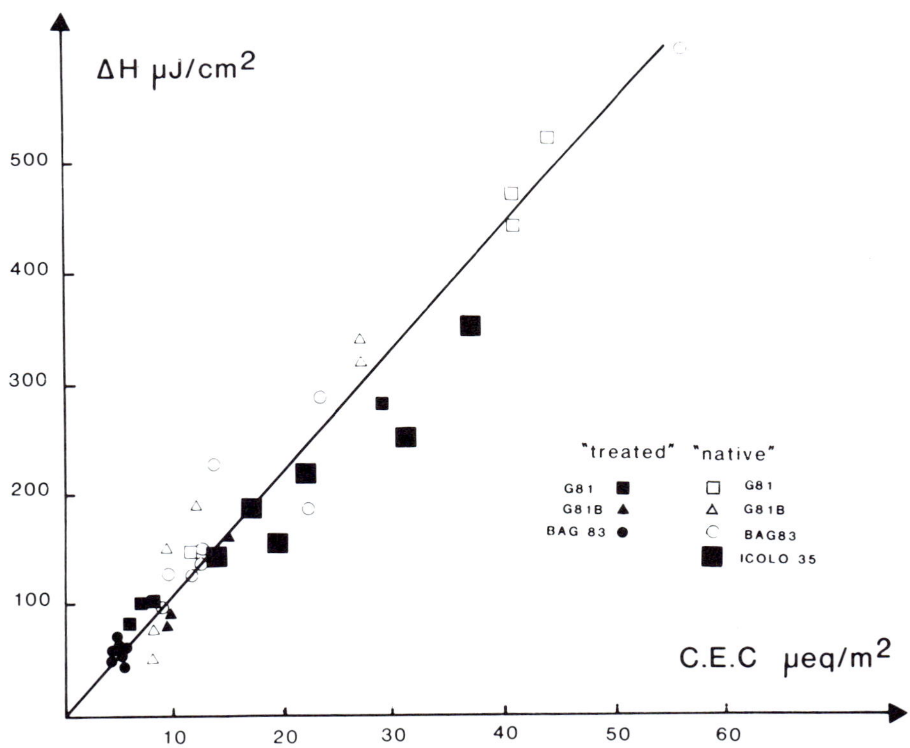

Fig. 13. Correlation between heat of immersion and cation exchange capacity in the Gironde and Loire estuaries. Data from years 1981 and 1983. All data from the Gironde estuary but ICOLO 35 (Loire estuary).

As for the other parameters, we performed measurements of heat of immersion on organics free and amorphous free samples. As in the case of CEC, the organic matter greatly contributes to the parameter and the marine O.M. is less active. This confirms the different reactivities of different types of organic matter, which we already observed for S.S.A..

In general, the absolute values of heat of wetting for both native and amorphous free samples are higher than for pure oxides or mineral (20-100 $\mu J/cm^2$, Zettlemoyer, 1965). However, for more basic oxides, having a higher pH_{zpc}, the heat of wetting becomes higher reaching 200 $\mu J/cm^2$ (Fubini, 1978). The typical range for organics free samples is 50-100 $\mu J/cm^2$, higher values being found for sea end members samples. Native and amorphous free samples show approximately the same heats of immersion, this similitude clearly indicates that Δ H is basically governed by organic coatings.

CONCLUSION

The role of organic matter and amorphous oxides has long been considered as a key to understand the geochemical interactions between dissolved and particulate phases. Indeed, this influence must be carefully assessed and greatly depends upon the considered property.

As far as electrophoretic mobility is concerned, we previously underlined the obvious lack of interaction between humic acids and negatively charged silica particles, however the same kind of experiments with model particles of higher isoelectric point as for example Al_2O_3 and iron oxides, indicate a strong effect of organics. The numerous studies performed by Liss and co workers showed that the electrophoretic mobility of both natural or synthetic particles was greatly influenced by fresh marine organic matter. There are strong evidences that a negatively charged organic film coats most of the natural particles.

Organic matter deposited onto the suspended sediment has a large influence on the specific surface area. On one hand, it appears that organic matter blocks some sites available for physical adsorption of inert gas (BET adsorption) and, on the other hand, it probably partly causes flocculation and agglomeration of particles in the upper estuary, such flocs and agregates being destroyed downstream in the salinity intrusion zone.

Two other properties, CEC and heat of immersion are strongly related. The comparison of the influence of coatings upon the values of these different parameters allows the distinction of two categories :

a) on the one hand, S.S.A. seems to be controlled by the amorphous oxides and is related to the Al content of the samples (which can be considered as an index of the fine grained fractions mostly composed of alumino-silicates), and accordingly to its granulometry, suggesting an almost uniform coating of the detritic core by amorphous oxides ; conversly, P.O.M. has a negative influence on this parameter.

b) on the other hand, C.E.C. and Heat of immersion which appear to be strongly correlated are controlled by both organic matter and amorphous oxides. CEC has long been used as a basic parameter to model the exchange of metals at the particulate surfaces.

Former studies by Hodgson et al. (1963) showed that the removal of organic matter, which partly controls CEC, from soils might either increase or decrease adsorption of Cobalt. It is likely that several parameters which are commonly associated with the adsorptive properties of particulate matter do not depend the same way on the constitutive properties of the samples. Nevertheless, it is still difficult :

i) to determine the main factor controlling adsorption (amorphous or organic coatings) keeping in mind that it is not necessarily the same for all metals,

ii) to determine the best parameter to estimate the adsorptive properties of particulate matter.

ACKNOWLEDGMENTS

The authors are indebted to the group for Electrophoresis of the Center for Marine Research Zagreb, "Rudjer Boskovic" Institute, Zagreb, Croatia, Yugoslavia, for CEC measurements. This study has been carried out with the financial support of C.N.R.S. (GRECO Interactions Continent-Ocean).

REFERENCES

Ahmed, S.M., 1966. Studies of the dissociation of oxide surfaces at the liquid-solid interface. Canc. J. Chem., 44 : 1163-1170.

Allen, G.P., 1972. Etude des processus sédimentaires dans l'estuaire de la Gironde. Thèse de Doctorat d'Etat. Université de Bordeaux I, France, 314 pp.

Barbaroux, L.H., 1980. Evolution des propriétés physiques et chimiques des sédiments dans le passage continent-océan. L'effet estuarien. (Estuaire de la Loire et de ses parages). Thèse de Doctorat d'Etat, Université de Nantes, France. 433 pp.

Benjamin, M.M. and Leckie, J.O., 1980. Adsorption of metals at oxide interfaces : effects of the concentrations of adsorbate and competing metals. In: R.A. Baker (ed.), Contaminants and Sediments. Ann Arbor Pub, Vol. 2.

Bolt, G.M., 1957. Determination of the charge density of Silica sols. J. Phys. Chem., 61: 1166-1169.

Busenberg, E. and Clemency, C.V., 1973. Determination of the Cation Exchange Capacity of clays and soils using an ammonia electrode. Clays, Clay Miner, 21, 213.

Davis, J.A. and Leckie, J.O., 1978. Surface ionization and complexation at the oxide/water interface. Part II. J. of Colloid and Interface Science, 67: 90-107.

Elbaz-Poulichet, F., Huang, W.W., Jednacak-Biscan, J., Martin, J.M. and Thomas, A.J., 1982. Trace metal behaviour in the Gironde estuary : the problem revisited. Thalassia Jugoslavica, 18: 61-96.

Elbaz-Poulichet, F., Holliger, P., Huang, W.W. and Martin, J.M., 1984. Lead cycling in estuaries, illustrated by the Gironde estuary, France. Nature, 308: 409-414.

Foissy, A., M'Pandou, A., Lamarche, J.M. and Jaffrezeic-Renault, N., 1982. Surface and diffuse-layer charge of the TiO_2-electrolyte interface. Colloids and Surf., 5: 363-368.

Fubini, B., Della Gata, G. and Venturello, G., 1978. Energetics of adsorption in alumina-water systems. Microcalorimetric study on the influence of adsorption temperature on surface processes. J. of Colloid and Interface Sci., 64: 470.

Hashimoto, I. and Jackson, M.L., 1960. Rapid dissolution of allophane and Kaolinite-Halloysite after dehydration. Clay and Clay Minerals, 7th Conf., pp. 102-113.

Haussmann, D.D. and Anderson, M.A., 1985. Using electrophoresis in modelling sulfate, salinity and phosphate adsorption on to goethite. Environmental Science and Technology, 19: 544-551.

Hunt, C.D., 1981. Regulation of sedimentary cation exchange capacity by organic matter. Chemical Geology, 34: 131.

Hunter, K.A., 1980. Microelectrophoretic properties of natural surface-active organic matter in coastal sea water. Limnol. Oceanogr., 25: 807-822.

Hunter, R.J., 1981. Zeta potential in colloid science. Academic Press, 386 pp.

Hunter, K.A. and Liss, P.S., 1982. Organic matter and the surface charge of suspended particles in estuarine water. Limnol. Oceanogr., 27: 322-335.

Hunter, K.A. and Liss, P.S., 1979. The surface charge of suspended particles in estuarine and coastal water. Nature, 282: 823-825.

Hunter, R.J. and Wright, H.J.L., 1971. The dependance of electrokinetic potential on concentration of electrolyte. J. of Colloid and Interface Sci., 37, n° 3: 564-580.

Jackson, O.P. and Mehra, M.L., 1960. Iron oxide removal from soils and clays by a dithionite-citrate system buffered with sodium bicarbonate. Clay and Clay minerals, 7th Conf., 317.

Juracic, M. and Pravdic, V., 1981. Geochemical and physico-chemical studies on sediments of the Rijeka bay. The properties of sediments as depositories of pollutants. Thalassia Jugoslavica, 17: 339-349.

Loder, T.C. and Liss, P.S., 1982. The role of organic matter in determining the surface charge of suspended particles in estuarine and ocean waters. Thallasia Jugoslavica, 18: 433-448.

Loder, T.C. and Liss, P.S., 1985. - Control by organic coatings of the surface charge of estuarine suspended particles. Limnology and Oceanography, 30: 418-421.

Martin, J.M., Jednacak, J. and Pravdic, V., 1971. The physico-chemical aspects of trace element behaviour in marine environments. Thalassia Jugoslavica, 7: 619-637.

Martin, J.M., Mouchel, J.M. and Thomas, A.J., 1984. Time concepts in hydrodynamic systems : an application to the Gironde estuary. Invited conference paper : VIII Int. Symp.: "Chemistry of the Mediterranean". Residence time of microconstituents in coastal waters, Primosten, Yugoslavia, May 16-24, Marine Chemistry (in press).

Morimoto, T., Nagao, M. and Omori, T., 1969. Heat of immersion of titanium dioxyde in water. I. Effect of the hydratation treatment of Titanium dioxide. Bull. Chem. Soc. Japan, 42: 943.

Neihof, R.A. and Loeb, G.I., 1972. Surface charge characterization of particulate matter in seawater. Limnol. Oceanogr., 17: 7-16.

Neihof, R.A. and Loeb, G.I., 1974. Dissolved organic matter in seawater and the electric charge of immersed surfaces. J. Mar. Res., 32: 5-12.

O'Connor, T.P. and Kester, D.R., 1975. Adsorption of Cobalt from fresh and marine systems. Geoch. Cosmoch. Acta, 39: 1531-1543.

Parfitt, G.D. and Sing, K.S.W. (eds), 1976. Characterization of powder surfaces. Academic Press, London, 476 pp.

Pravdic, V., 1970. Surface charge characterization of sea sediments. Limnol. Oceanogr., 15: 230-235.

Pravdic, V., Jednacak-Biscan, J. and Juracic, M., 1981. Physico-chemical parameters describing the role of particulate materials in estuarine waters in River Inputs to Ocean Systems. J.M. Martin, D. Eisma and J.D. Burton (eds). UNEP and UNESCO, pp. 188-196.

Smith, A.L., 1976. Electrokinetics at the oxide-solution interface. J. of Colloid and Interface Science, 55: 525-530.

Tiller, K.G., Hodgson, J.F. and Peech, M., 1963. Specific sorption of Cobalt by soil clays. Soil Science, 95: 392-399.

Tipping, E., 1981. The adsorption of humic substances by ion oxides. Geoch. Cosmoch. Acta, 45: 191-199.

Tiselius, A., 1937. A new apparatus for electrophoretic analysis of colloidal mixtures. Trans. Faraday Soc., 33: 524-531.

Tiselius, A., 1938. Elektrophoretische Messungen am Eiweiss. Kolloid Z., 55: 129-137.

Zettlemoyer, A.C., 1965. Immersional wetting of solid surfaces. Industrial and Engng. Chemistry, 57: 27-36.

II

NUTRIENT CYCLING AND PATHWAYS OF ORGANIC TRANSFORMATIONS

MODELLING STUDIES OF MATERIAL FLOWS IN A SHALLOW ECOSYSTEM COMPARED TO THE OPEN OCEAN

J.G. FIELD, P.A. WICKENS and C.L. MOLONEY

Marine Biology Research Institute, Zoology Department, University of Cape Town, 7700 Rondebosch (South Africa)

ABSTRACT

Field, J.G., Wickens, P.A. and Moloney, D.L., 1986. Modelling studies of material flows in a shallow ecosystem compared to the open ocean. In: P. Lasserre and J.M. Martin (eds), Biogeochemical Processes at the Land-Sea Boundary. Elsevier, Amsterdam.

Two nitrogen flow simulation models are presented to contrast the effects of horizontal water transport which is believed to be important in shallow-water marine systems,with vertical transport in an open ocean plankton community, including the microplankton. A kelp-bed community dominated by filter feeders is modelled as a closed system with no transport, and under both upwelling and downwelling conditions. In the closed system, feedback of organic nitrogen through detritus formation from animal faeces is important in the maintenance of a food supply, a situation analogous to closed bays and estuaries.

Under upwelling conditions, surface water transports material out of the system and cool, clear upwelling water dilutes the suspension of detritus. Under downwelling conditions when phytoplankton-rich water from offshore is imported, phytoplankton supplements the detrital diet of shallow-water filter-feeders and their growth is faster ; this situation parallels that found on exposed rocky intertidal shores.

A nitrogen flow model of a microplankton community is used to investigate the effect of sedimentation out of the mixed layer. The simulation shows a series of peaks of phytoplankton, bacteria, zooflagellates, larger protozoa and micro-and meso-zooplankton over a 13-day period after an upwelling or mixing event. Two phytoplankton peaks develop, corresponding to "new" and "regenerated" production respectively. Increasing rates of sedimentation of phytoplankton and zooplankton faeces out of the mixed layer result in slightly slower development of the bloom, flattened and smaller peaks of all components, increasing relative importance of "regenerated" production, and maintenance of a larger phytoplankton P/B ratio. The series of peaks in the model suggests that different components of the community may dominate the biomass at different successional stages in the development of a bloom. This may help to resolve the controversy over the significance of micro-organisms in the plankton community.

INTRODUCTION

During the 1970's a good deal of research was done on energy flows through coastal ecosystems, tracing the flow of energy or carbon from primary producers to consumers. These studies included measurements of primary production, often expressed as mean annual values (Table 1). It can be seen from Table 1 that the highest primary production tends to occur in systems with a large proportion of macrophytes, regardless of latitude.

Presumably the high productivity is due to attached macrophytes maintaining position and exposure to sunlight in the face of tidal, wind- or wave-induced water movement. This allows recycling of minerals from the sea floor to the plants, in contrast to micro-algae on the sediments and phytoplankton which are likely to be transported and/or shielded from light by turbid waters when there is water movement at shallow depths.

TABLE 1

Some primary production estimates in coastal ecosystems. The values of primary production have been compiled from the sources indicated and percentages of primary production allocated to three categories of primary producer.

SYSTEM	$gC.m^{-2}yr^{-1}$	% Macrophytes	% Attached Micro-algae	% Phytoplankton
(1) Kelp-bed (Nova Scotia)	1976	89 %	-	11 %
(2) Salt-marsh (Sapelo Island)	1445	84 %	10 %	6 %
(3) Kelp-bed (Benguela)	1270	60 %	-	40 %
(4) Mangal (Florida)	890	100 %	-	-
(5) Enclosed Bay (Narragansett)	400	-	-	100 %
(6) North Sea	320	-	-	100 %
(7) Baltic Sea	266	28 %	-	72 %
(8) Mud flat (Cornwall, U.K.)	225	-	64 %	36 %
(9) Tropical Estuary (Sierra Leone)	186	-	-	100 %

Authors : (1) Miller et al. (1971) ; (2) Pomeroy & Weigert (1981) ; (3) Newell et al. (1982) ; (4) Lugo et al. (1978) ; (5) Nixon et al. (this volume) ; (6) Joiris et al. (1982) ; (7) Jansson et al. (1982) ; (8) Warwick et al. (1979) ; (9) Longhurst (1983).

Many of the ecosystem studies referred to in Table 1 have been summarised by descriptive models of mean annual energy flow, which have been useful for quantifying the average flows in food webs involved in such ecosystems. They have also led to studies on the relative importance of detritus in coastal ecosystems (see Man, this volume). If water movement is as important for coastal systems as suggested above, then its effect should be studied. This has been done in some instances ; in the Baltic Sea, for example, Jansson et al. (1982) have demonstrated that water movements and fish migration provide the links between pelagic, bottom and littoral subsystems transporting energy and materials to maintain nutrient cycling in the whole system. Smetacek (1984) has related benthic processes to those in the water column above and, similarly, Joiris et al. (1982)

and Nixon (this volume) have shown that the transport of material between the water column and sediments and back maintains the productivity of both subsystems.

In a review of Benguela kelp-beds on the west coast of South Africa, Newell et al. (1982) presented an annual budget of primary production and energy requirements by consumers, summarising detailed analyses of biomass, distribution and eco-physiological studies of key filter-feeders such as the ribbed mussel Aulacomya ater, and predators such as the commercially exploited rock lobster Jasus lalandii.

The balance between primary production and consummer requirements in that budget might appear to suggest that the kelp beds form a closed system, or alternatively, that imports balance exports. Measurements of water movements through the kelp beds (Field et al., 1980b) indicate that wind-driven upwelling may result in the water column turning over up to seven times per day, with detritus-laden water being exported at the surface and clear, phytoplankton-free water from below the euphotic zone replacing it from beneath.

In contrast, under downwelling conditions phytoplankton blooms which develop offshore in warmed plumes of previously upwelled water, move back towards the coast and enter the fringing kelp-beds at the surface, displacing denser water beneath. Thus under downwelling conditions, filter-feeders are provided with a rich supply of imported phytoplankton. The system is therefore essentially an open one, with imports and exports governed by water transport. It is extremely difficult to measure the effect of a dynamic physical process such as water transport on the rate at which an ecological community functions, but hypotheses can be tested for consistency using simulation models as has been done for a Benguela kelp-bed by Wulff and Field (1983).

In the deeper ocean, phytoplankton primary production values are generally considerably lower than those given for coastal waters in Table 1 (Ryther, 1969 ; Pace et al., 1984), with the obvious exception of upwelling systems such as the Peruvian (Walsh et al., 1981) and Benguela (Brown, 1984 ; Shannon and Field, 1985) systems. The high productivity of upwelling systems is ascribed to vertical transport of nutrient-rich water into the photic zone, and this is often accompanied by downward flux of organic matter after phytoplankton blooms in upwelling areas (Walsh et al., 1981). In the Celtic Sea sedimentation of phytoplankton has been observed as it falls to the bottom (Billett et al., 1983). Thus, in the open ocean vertical flows are identifiable and appear to be important, whereas they tend to be combined with wave mixing or horizontal transport in shallow marine systems.

It is the aim of this contribution to contrast the effect of vertical fluxes of material in the open ocean with horizontal transport which is likely to be important in shallow-water systems. The models described below are based on a number of simplifying assumptions, therefore the results must be treated with caution. In particular the magnitude of predictions may not always be realistic, although the relative proportions and sequence of events modelled are believed to reflect the real situations.

NITROGEN FLOW IN A BENGUELA KELP BED

Velimorov et al. (1977) have described Benguela kelp bed communities which are subjected to pulsing water movement due to alternating upwelling and downwelling conditions in the Benguela current (Field et al., 1980 b) and thus pulsing availability of phytoplankton to filter-feeders. The primary producers in the kelp bed are phytoplankton and macrophytes. The dominant macrophytes are Laminaria pallida and Ecklonia maxima, with occasional Macrocystis angustifolia and various understorey algae and kelp epiphytes described by Field et al. (1980 a). The phytoplankton and macrophytes yield 86.5 g N m^{-2} yr^{-1} and 58.8 g N m^{-2} yr^{-1} respectively (Newell and Field, 1983 a). The production by macrophytes is released as particulate organic matter (POM) and dissolved organic matter (DOM) in carbon : nitrogen proportions of 9:1. The nitrogen component of primary production by phytoplankton constitutes 68 % of the total, while in terms of carbon, phytoplankton only constitutes 35 % of all primary production. Macrophytes have a C:N ratio approximately three times that of phytoplankton (Newell and Field, 1983 a) which accounts for phytoplankton being the more important primary producer in terms of nitrogen. A characteristic feature of the kelp bed community is the scarcity of grazers directly utilizing the living kelp, as they only constitute some 4% of the total kelp consumer community (Velimirov et al., 1977 ; Field et al., 1977 and Newell et al., 1982). Therefore, most of the macrophyte production passes along the detrital food chain to the filter-feeders which form 72 % of the total faunal standing stock in the community. Of this dominant community, 46 % is the mussel, Aulacomya ater, 13 % sponges (Tehya spp.), 6 % ascidians (Pyura spp.) and the remaining 36 % a variety of other small filter-feeders. Carnivores include the rock lobster, Jasus lallandii, anemones, the fish, Pachymetopan blochii and the isopod, Cirolana imposita (Shafir and Field, 1980).

Bacteria and faeces are components of the microbial loop in which detrital material is degraded by bacterial action into a form that can be utilized by the filter-feeders. Newell (1965) showed that bacteria can colonise faecal material causing an increase in faecal nitrogen content, enhancing its nutritional value. Rice (1982), however, found that bacteria accounted for very little of the nitrogen and the percentage increase in nitrogen was largely due to aging of the detritus itself. There are two factors which make bacteria and faeces important components. Firstly, particles can be refiltered numerous times, creating cyclic flows through the system. Secondly, bacteria are significant in converting dissolved organic matter into particulate form for consumption by filter-feeders which cannot utilize dissolved matter directly.

Kelp bed model.

Data used for the model are based on a hypothetical kelp bed in the southern Benguela region stretching for 10 km of coast with a mean width of 500 m extending to 20 m depth, with a mean depth of 10 m as in Newell et al. (1982). All units are expressed per square metre.

The kelp-bed system may be depicted in terms of flows of energy, carbon or nitrogen, although in this model the currency is nitrogen, which Newell and Field (1983 b) found to be a potentially limiting nutrient in this system. The flow pathways show the feedback loops provided by animal faeces contributing to detritus formed of kelp material and bacteria shown separately, as indicated in Figure 1.

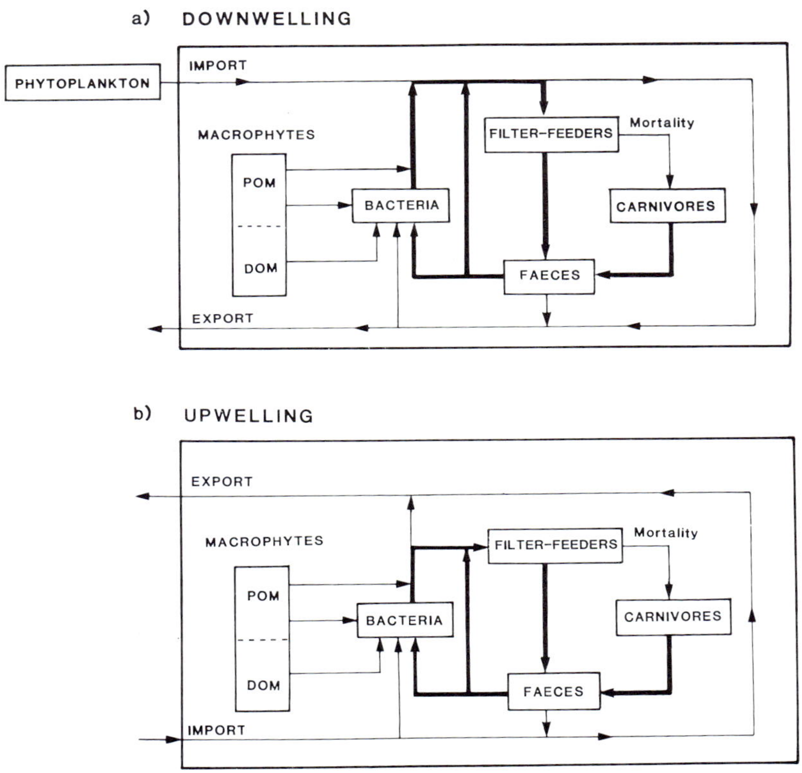

Fig. 1. Model depicting nitrogen flows in a kelp bed community. Primary production by macrophytes is partitioned into particulate (POM) and dissolved (DOM) components. Filter-feeders feed on detritus consisting of POM, bacteria and animal faeces. Recycling of nitrogen via the feedback loop provided by faeces is indicated by heavy lines. Fig. 1a) shows the model under downwelling conditions, when phytoplankton is imported with surface water from offshore. Fig. 1b) shows the model under upwelling conditions when it is assumed that phytoplankton in the upwelling water is negligible and excess detritus is exported in surface water.

The system includes macrophytes (kelp, understorey and epiphytic algae), bacteria, filter-feeders, carnivores and animal faeces. Grazers are negligible in biomass and omitted from the model. Phytoplankton is regarded as being outside the system and both macrophyte and phytoplankton primary production are assumed for simplicity to be constant. Nitrogen flow through the system originates as macrophyte primary production in the form of particulate and dissolved organic matter (POM and DOM, respectively).

Phytoplankton primary production is available in the kelp-bed under downwelling conditions (Fig. 1a) when it is regarded as being imported even if there is no flow ("downwelling with zero transport"). Under upwelling conditions no phytoplankton is included in the model (Fig. 1b), since cool upwelling water is normally clear and contains negligible phytoplankton (Field et al., 1980b).

Nitrogen is lost from the system through water exchange and carnivore mortality. For ease of description, the primary producers (phytoplankton, POM, DOM), bacteria and faeces will be referred to as food components in contrast to the filter-feeders and carnivores which will be referred to as consumer components. The sequence of computations for the food components at each time step is : (1) water exchange (loss/gain); (2) filter-feeder consumption (loss) ; (3) bacterial degradation (loss) ; and (4) primary production (gain). Details of equations are given in Wickens and Field (1985).

The changes of the standing stocks of food components at the end of each day are calculated as :

Final standing stock = initial standing stock ± losses/gains due to water transport - losses to filter-feeders - losses to bacteria + production (1)

Water exchange affects the standing stock of each food component and depends on the difference between the external and internal standing stocks. The loss/gain through net transport of the component is a hyperbolic function of the exchange rate, taking into account the time step.

Net transport = (1 - $\frac{1}{2}$exchange rate/time step) x (internal - external standing stocks) (2)

The filter-feeders are assumed to filter constant proportions of phytoplankton, bacteria, faeces and POM from the water column, so that losses to filter-feeders are calculated for each component as :

Food filtered = Standing stock x electivity index (3)

Bacterial break-down of the remaining phytoplankton, faeces and POM occurs after the filter-feeders have filtered out their ration. In addition the dissolved organic matter (DOM), which cannot be utilised directly by filter-feeders, is converted to bacterial production. Thus bacterial consumption of each component is represented by :

Bacterial consumption = Food available x degradation rate (4)

Bacterial production from each food component is calculated by multiplying consumption by bacterial nitrogen conversion efficiencies, and these are then summed to calculate the total daily bacterial production. The faeces component of the model includes both faeces and urine of filter-feeders and carnivores. Total faecal production of both is therefore calculated as the total amount of food consumed by filter-feeders and carnivores minus that assimilated, to which must be added the amounts of nitrogen excreted. Macrophyte and phytoplankton primary production are assumed for simplicity to be constant, but phytoplankton primary production is only available in the kelp-bed under downwelling conditions.

The consumer standing stocks (filter-feeders and carnivores) change according to the balance between net growth and motality. This is calculated as follows :

Final standing stock = inital standing stock + food assimilated
- nitrogen excreted - mortality (5)

The initial and final standing stocks are those calculated for the beginning and end of one day, respectively. An assimilation constant for each of the filtered food components is used to calculate the amount of food that is assimilated into filter-feeder standing stock. The biomass of filter-feeders that are consumed by carnivores is calculated as :

Filter-feeder mortality = filter-feeder biomass x carnivore consumption
x carnivore biomass (6)

Carnivore mortality is calculated as :

Carnivore mortality = (carnivore biomass)^2 x constant (7)

thus allowing for curvilinear density dependence. Excretory losses for both filter-feeders and carnivores are assumed to be constant proportions of food consumed.

Standing stocks. The nitrogen equivalents of biomass of phytoplankton, bacteria, filter-feeders and carnivores were converted from calorific values, as detailed by Wickens and Field (1985). Filter-feeders produce an estimated 113 g N m^{-2} y^{-1} of faeces and urine combined (Newell and Field, 1983 b), with a corresponding figure of 15,3 g N m^{-2} y^{-1} for carnivores (Newell and Field, 1983a). Summing these and converting them to daily production, a value of 0.35 gN m^{-2} d^{-1} for faeces was used in units of gN m^{-2} to start each simulation. The calculated values of standing stocks of faeces and POM are confirmed by those measured in southern Benguela kelp beds by Stuart and Klumpp (1984).

Standing stocks of the food components in water outside the kelp-bed system were set to zero under upwelling conditions when cool, clear water enters the kelp bed from below the photic zone. Under downwelling conditions phytoplankton becomes an important component of the suspended matter, and the value of 0.237 gN m^{-2} was used, equivalent to the mean daily phytoplankton production, calculated from Newell and Field (1983 a).

Primary production. The yearly production values for the primary producers were obtained from Newell and Field (1983 a), and converted to daily production values. During upwelling conditions when it is assumed that no phytoplankton is present in the kelp bed and none is imported, production is set to zero, whereas under downwelling conditions, daily production was calculated to be 0.237 gN m^{-2} d^{-1}.

Biological variables. Electivity indices (or filtration efficiencies) for the four suspended food components (1 ; 0.8 ; 0.7 ; 0.7 for phytoplankton, bacteria, faeces and POM respectively) were the values found to be most realistic by Wulff and Field (1983) based on eco-physiological work on kelp bed mussels (Griffiths and King, 1979a and b, and Stuart, 1982). Carbon assimilation efficiencies of mussels have been found to be 40-50 % for kelp POM, 50 % for detritus and 67-70 % for bacteria (Stuart et al., 1982 a, b). No measurements have been made of the corresponding nitrogen assimilation efficiencies, but calculations indicate that they are likely to be higher than for carbon (Newell and Field, 1983b), and values of 0.8 ; 0.8 ; 0.6 ; 0.5 were assumed for phytoplankton, bacteria, POM and faeces respectively. The sensitivity the model to these assumptions was checked by means of sensitivity analyses.

Bacterial degradation rates of components were given the values as calculated for carbon by Wulff and Field (1983), namely 0.033 % d^{-1} for phytoplankton, POM and faeces and 0.1 % d^{-1} of DOM are degrated by microbial action. Bacterial carbon conversion efficiencies were shown to be between 10 % (Newell and Lucas, 1981) during downwelling, and a possible 40 % (Linley and Newell, 1985) during upwelling conditions. Considerably higher nitrogen conversion efficiencies (NC) have been measured for bacteria such as 94.2% for conversion of kelp POM (Koop et al., 1982) and 83 % for faeces (Stuart et al., 1982a). In order to obtain a single value for nitrogen, the proportion of faeces concentration to kelp concentration was calculated and each of the efficiency values weighted accordingly. The ratio of suspended faeces to kelp POM is 24:1, giving a net conversion efficiency of 85 %.

Filter feeder- and carnivore- related constants were calculated from the literature but they were altered whilst tuning the model (see Wickens and Field, 1985 for details). The final tuning was done on the carnivore mortality rate. In addition, the standing stocks of the filter-feeders and carnivores were initiated to values i) below and ii) above their calculated values, in a closed system with no water transport, and tuned to reach their original values after several years. Thus the model is tuned to maintain stability of the primary and secondary consumers in a closed system. The model was programmed in GBASIC for an Apple II Computer.

KELP-BED MODEL OUTPUT

The nitrogen flow model was tested for sensitivity to varying the "biological parameters" of filter-feeder electivity and assimilation efficiency. Results detailed in Wickens & Field (1985) show that the model is not sensitive to varying biological parameters when the system is closed (i.e. no water transport through upwelling or downwelling), and that the model is most sensitive to varying feeding efficiencies under upwelling conditions when suspended matter is transported by water movement. A comparable simulation model of energy flow through a Benguela kelp-bed (Wulff & Field, 1983) showed that water exchange on the scale measured in the field by Field et al. (1980b) has a far greater effect on the potential growth rate of filter feeders than realistic ranges of biological variables such as feeding efficiencies.

Fig. 2 depicts the relative proportions of different suspended food components available to filter-feeders with different rates of upwelling (transporting surface water out of the system) and downwelling (transporting surface water horizontally into the nearshore system). When modelled as a "closed system" (i.e. water transport) bacteria (including other micro-organisms) contribute to the nitrogen standing stock of suspended material. Under these circumstances recycling of faeces is an important process.

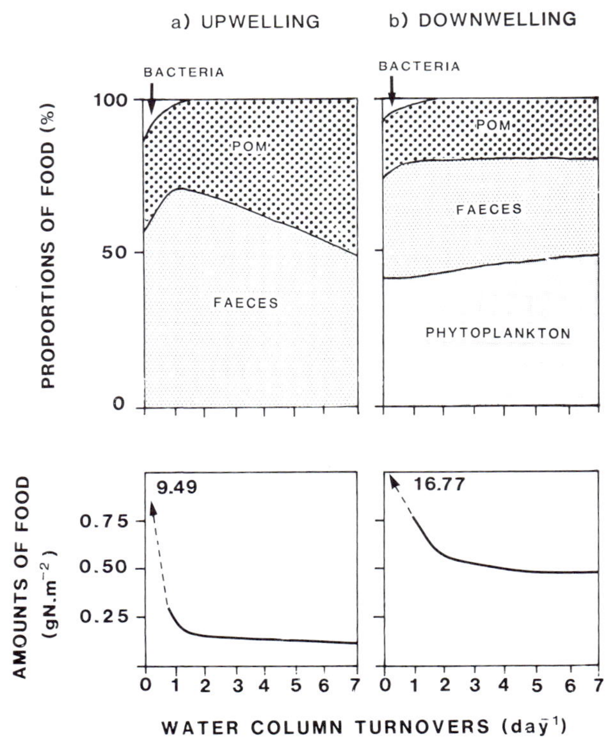

Fig. 2. Model output showing the percentage of different food components available to filter feeders with varying water transport (0-7 water column turnovers per day). The nitrogen content of all food available to filter feeders is shown below, and is much higher when the system is "closed" (0 turnovers). Fig. 2a) shows food proportions and quantities under upwelling conditions when all food is derived from macrophytes. Fig. 2b) depicts downwelling conditions when phytoplankton is an additional component (After Wickens and Field, 1985).

There are two categories of "closed system" : under upwelling conditions but with no water transport, it is assumed that upwelling has recently ceased leaving clear water in the system with negligible phytoplankton content and the only source of primary production is derived from macrophytes. On the other hand, under downwelling conditions with no water transport, phytoplankton in the kelp-bed water column make an additional contribution to the primary production and the total supply of food to filter-feeders is greater. Thus one may generalise that in closed nearshore systems, detritus and faeces cycling are likely to be more important than in open systems, and that micro-organisms attached to faeces and other detritus enhance its food value to filter-feeders. The closed kelp bed is probably analagous to closed bays and closed estuaries.

Under conditions of continuous downwelling, phytoplankton imported from outside the system becomes increasingly important with faster rates of water transport (Fig. 2b) and at rates faster than one water-column turnover per day, phytoplankton contribute more nitrogen to filter-feeder food than macrophyte particulate matter or recycled faeces.

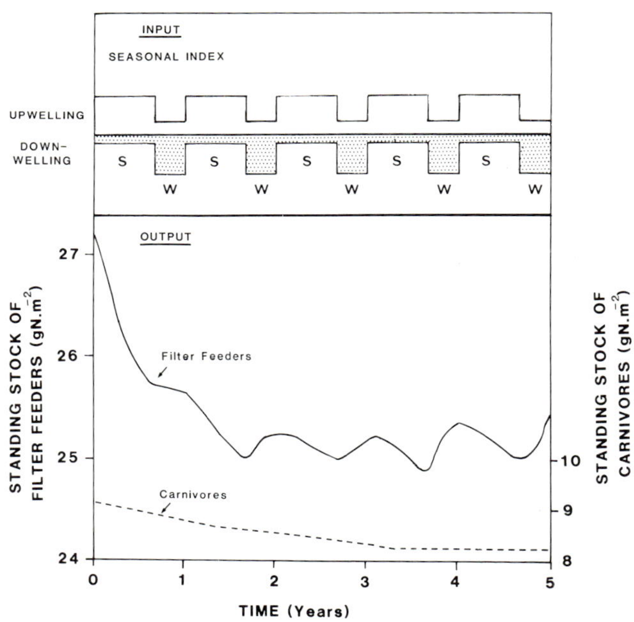

Fig. 3. Input and output of kelp bed model when realistic pulses of upwelling and downwelling water transport are fed into the model. Upwelling prevails during summer and downwelling during winter. Model output (below) shows that the standing stock of filter feeders declines from the closed system level and then oscillates, increasing during winter and decreasing during the summer season. Carnivore standing stock declines when the system is pulsed (After Wickens and Field, 1985).

These conditions are analagous to exposed rocky shores in which the nearshore system is dominated by consumers which feed on phytoplankton imported from a vast reservoir of plankton over the continental shelf. Here the rate of transport of food into the nearshore system is as important as the concentration of plankton in determining the rate of food supply. When continuous upwelling is modelled, the proportion of particulate matter derived from macrophytes increases with faster water transport, and recycling of faeces and microbes both contribute less to suspended particulate nitrogen. The horizontal transport of water out of the nearshore system depletes detritus and faeces in the water column, and the clear replacing water is deficient in phytoplankton. Thus the food supply under upwelling conditions is paradoxically reduced. Such conditions of reduction in the nearshore food supply by offshore transport may be unique to upwelling regions, but there are analogies with estuarine flushing while rivers are in spate, and later enrichment when calm conditions promote primary production.

Under natural conditions, nearshore water transport is seldom continuous but fluctuates under the influence of winds and tides. In the southern Benguela region, the kelp beds are mainly influenced by the wind pattern which varies seasonally. Typically, the annual pattern can be simplified to eight months of spring-summer conditions and four months of winter conditions (Wulff & Field, 1983). The summer pattern is typified by repeated cycles consisting of four days of strong upwelling, two days of moderate upwelling, followed by four days of calm or onshore conditions when moderate downwelling occurs. In winter downwelling prevails, with a pattern of 18 days of moderate downwelling followed by two days of moderate upwelling. Sensitivity analyses of different combinations of pulse legth and rates of water transport show that the model is most sensitive to fast rates of upwelling and less sensitive to pulse duration, with longer pulses decreasing the food supply more than short pulses (Wickens & Field, 1985).

Fig. 3 shows the seasonal input of upwelling/downwelling described above, with the output of fluctuations in filter-feeder and carnivore nitrogen standing stocks predicted by the model. The filter-feeder standing stock declines from the closed system steady-state value, to fluctuate annually about a slightly lower level, increasing during the winter (onshore transport) season, and decreasing during the summer (offshore transport) season. In the absence of carnivores the fluctuation is much greater (not shown).

MODEL OF A PLANKTON BLOOM

Model of plankton blooms date back to the classical work of Riley et al. (1949), later advanced by Steele (1974). The role of microbial organisms was incorporated in models of the pelagic plankton by Pomeroy (1979), Pace et al. (1984) and Fasham (1985). The above works have all treated plankton communities on a seasonal scale. In the model described below, we deal with a simple model of phytoplankton and micro-organisms concerned with the "microbial loop" (Azam et al., 1983) on a daily time scale appropriate to the turnover time of these organisms. The model is based on biomass data collected by Holligan et al.

(1984a, b) and Newell and Linley (1984) in the English Channel in July/August 1981. Methods of collection are given in detail by Holligan et al. (1984a). We model the flows of nitrogen in a uniform mixed layer of the water column, with the bottom boundary of the model system at the thermocline.

The plankton community is divided into five components on the basis of function and size (Fig. 4). Primary producers are all grouped together in one phytoplankton compartment. The consumer compartements comprise bacteria, zooflagellates which prey on bacteria (Azam et al., 1983), larger non-photosynthetic protozoa which are assumed to feed on zooflagellates and phytoplankton, and micro- and meso-zooplankton (collected on 80 and 200 um mesh, respectively) which we assume to feed on both phytoplankton and large protozoa. Macro-zooplankton and fish are excluded from the model for lack of suitable data. For simplicity, the abiotic nitrogen pool is represented by a single compartment, with no distinction being made between urea, ammonia and nitrate.

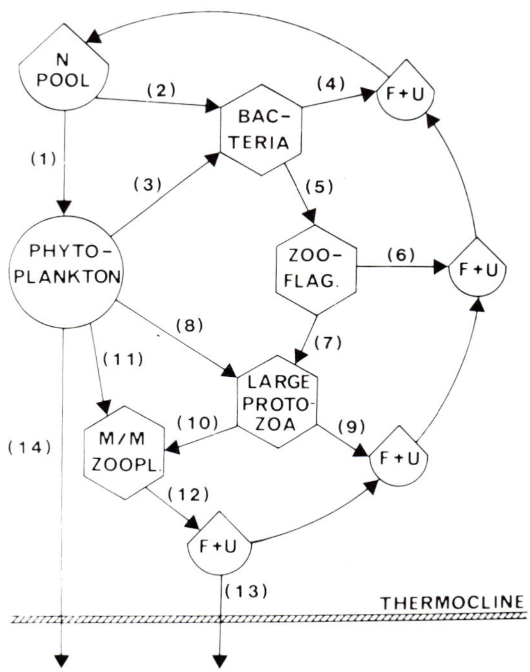

Fig. 4. Compartmental model describing the cycling of nitrogen in a planktonic community in the mixed layer of a water column. Flow pathways are represented by arrows and numbers which correspond to mathematical expressions described in Table 2. The nitrogen pool represents all abiotic nitrogen (nitrate, ammonia and urea), and other compartments represent bacteria, zooflagellates, larger protozoa, and micro-mesozooplankton, giving off waste products (F+U). Arrows (13) and (14) depict sedimentation of zooplankton faeces and phytoplankton cells, respectively (After Moloney et al., 1985).

The simulation model depicts the flows of nitrogen between the compartments, and in particular is used to investigate the effect of sedimentation of phytoplankton and zooplankton faeces out of the euphotic zone which is assumed to be 60 m deep ; however the depth does not affect the conclusions drawn from the results.

Nitrogen fluxes in the plankton model

The nitrogen flow pathways are shown in Fig. 4 and numbered from 1-14. All the flows are represented by one of three expressions :

Specific uptake rate = $k_1 \; Y \; (1 - e^{-k_2 \; X})$ (8)

where the rate of uptake from the donor compartment (X) to the recipient (Y) increases asymptotically to k_1 as X increases. The expression is similar to the Michaelis-Menten expression (see Moloney et al., 1985 for details). The asymptotic function was used to represent nitrogen fluxes from the abiotic pool to phytoplankton (Dugdale and Goering, 1967 ; Eppley et al., 1969 ; MacIsaac and Dugdale, 1972). It was also used to model the competitive uptake of abiotic nitrogen by bacteria (Monod, 1949), predation on bacteria by zooflagellates as described by Fenchel (1982), and the assumed manner of bacterial uptake of nitrogen from moribund phytoplankton cells.

TABLE 2

Expressions used in the plankton model described in Figure 4. Numbers in parentheses correspond to pathways numbered in Fig. 4. Equations below describe the rate of change of each compartment based on the expressions numbered above. Constants K_1 to K_{22} are given in Table 3. From Moloney et al. (1985).

(1) $K_1 \; P \; (1 - e^{-K_2 \; N})$	(8) $K_{14} \; P \; L$
(2) $K_3 \; B \; (1 - e^{-K_4 \; N})$	(9) $K_{15} \; L + K_{16} \; [(7) + (8)]$
(3) $K_5 \; B \; (1 - e^{-K_6 \; P})$	(10) $K_{17} \; L \; Z$
(4) $K_7 \; B + K_8 \; [(1) + (2)]$	(11) $K_{18} \; P \; Z$
(5) $K_9 \; F \; (1 - e^{-K_{10} \; B})$	(12) $K_{19} \; Z + K_{20} \; [(10) + (11)]$
(6) $K_{10} \; F + K_{12} \; [(5)]$	(13) $K_{21} \; [(12)]$
(7) $K_{11} \; F \; L$	(14) $K_{22} \; P$

dN/dt = (4) + (6) + (9) + (12) - (13) - (1) - (2)
dP/dt = (1) - (3) - (8) - (11) - (14)
dB/dt = (2) + (3) - (4) - (5)
dF/dt = (5) - (6) - (7)
dL/dt = (7) + (8) - (9) - (10)
dZ/dt = (10) + (11) - (12)

The second form of expression represents Lotka-Volterra interactions :

Interaction rate = k_3 X Y (9)

in which the donor compartment (X) is consumed by the recipient (Y) at a rate determined by k_3. The four pathways from phytoplankton to large protozoa and to micro/mesozooplankton, from large protozoa to micro/mesozooplankton and from zooflagellates to large protozoa have been modelled in this way.

The third type of expression represents release rates of nitrogen :

Release rate = k_4 Z (10)

which is used to describe the release rate of faeces and excreta from bacteria, zooflagellates, large protozoa and zooplankton compartments, where Z is the rate of nitrogen ingestion by the respective compartments.

TABLE 3

Values assigned to each of the 22 constants in the 14 expressions used in the simulation model, and the relevant sources. The flow pathways are represented as donor -> recipient compartments. N = nitrogen pool ; P = phytoplankton ; B = bacteria ; F = zooflagellates ; L = large protozoa ; Z = micro-mesozooplankton ; (F+U) = faeces and urine (nitrogen pool). All units expressed in terms of mg, m^2 and/or d.

Const.	Compartments	Value	Source
K_1	N -> P	0.9	Harrison et al. (1977)
K_2	N -> P	0.0025	Harrison et al. (1977)
K_3	N -> B	0.9	By comparison with K_1
K_4	N -> B	0.0462	By comparison with K_1
K_5	P -> B	4.5	Indirectly from Newell and Linley (1984)
K_6	P -> B	0.0008	Indirectly from Newell and Linely (1984)
K_7	B -> (F+U)	0.07	By comparison with carbon respiratory losses
K_8	B -> (F+U)	0.63	Newell and Linley (1984)
K_9	B -> F	17	Fenchel (1982)
K_{10}	B -> F	0.0023	Fenchel (1982)
K_{11}	F -> (F+U)	0.05	By comparison with carbon respiratory losses
K_{12}	F -> (F+U)	0.65	Newell and Linley (1984)
K_{13}	F -> L	0.0043	Using data from Newell and Linley (1984)
K_{14}	P -> L	0.026	Using data from Newell and Linley (1984)
K_{15}	L -> (F+U)	0.05	By comparison with carbon respiratory losses
K_{16}	L -> (F+U)	0.65	Newell and Linley (1984)
K_{17}	L -> Z	0.002	Assumed equal to K_{18}
K_{18}	P -> Z	0.002	Using data from Newell and Linley (1984)
K_{19}	Z -> (F+U)	0.05	By comparison with carbon respiratory losses
K_{20}	Z -> (F+U)	0.40	Newell and Linley (1984)
K_{21}	Z(F+U) -> SED	0 - 0.3	
K_{22}	P -> SED	0 - 0.3	

The nitrogen released serves as input to the abiotic nitrogen pool as shown in Fig. 4. In addition, a small starvation rate of 5 % of the standing stock per day was assumed to be excreted in the absence of food.

The equations used in the model are listed in Table 2 and described in more detail in Moloney et al. (1985). Table 3 gives values for the constants employed in the equations, together with the flow pathways and the sources of information used to obtain the values. The main purpose of the model is to investigate the effect of vertical transport out of the mixed layer by sedimentation. The two main compartments that may sustain such losses are phytoplankton and zooplankton faeces (Joint and Morris, 1982 ; Fasham, 1985). In general, phytoplankton sinking rates (Smayda, 1970 ; Bienfang and Harrison, 1984) are slower than those of zooplankton faeces (Joint and Morris, 1982), and Honjo (1978) has suggested that most of the vertical flux in the water column is due to faecal pellets (however, see also Billett et al., 1983). The rate of sedimentation is modelled simplistically as a constant proportion of the total phytoplankton standing stock and the zooplankton faeces, the proportion being varied in different simulations. The set of differential equations comprising the model are solved numerically by a second-order Runge-Kutta approximation, using a time step of 0.05 day.

Sensitivity analyses showed that the model displayed the same general behaviour and trends when the constants used in the equations were varied up and down five-fold (Moloney et al., 1985). However, the model is sensitive to starting values of nitrogen standing stocks in each compartment. For simplicity, all starting values of living compartments were set to low values with realistic relative proportions, whereas the simulations were initiated with an abiotic nitrogen pool equivalent to 1.8 μg.at.N.l^{-1} to simulate an upwelling or mixing event with water from below the thermocline.

PLANKTON MODEL OUTPUT

Fig. 5 shows the levels of nitrogen standing stock in each compartment during a series of 13-day simulations. The simulations all commence with a large upwelling or mixing event in which nitrogen-rich water with very small standing stocks of phytoplankton, microbes and zooplankton is mixed into the photic zone from below the pycnocline. In Fig. 5A the system is modelled without any sedimentation of faeces or phytoplankton out of the mixed layer. The abiotic nitrogen pool, which initially would consist largely of nitrate-N, declines rapidly as a phytoplankton bloom develops and peaks after about 4.5 days, corresponding with "new production" (Dugdale and Goering, 1967). The subsequent decline in phytoplankton biomass is due to the combined effects of grazing by herbivores and depletion of the abiotic nitrogen pool by both phytoplankton and bacteria. The bacteria reach a peak after some 5.3 days, followed by zooflagellates (5.7 days), and larger heterotrophic protozoa (7 days). The phytoplankton standing stock recovers after 6.5 days when the zooplankton have begun to increase. Zooplankton reach maximum biomass after about 9 days and thereafter remain fairly dominant ; the zooplankton development

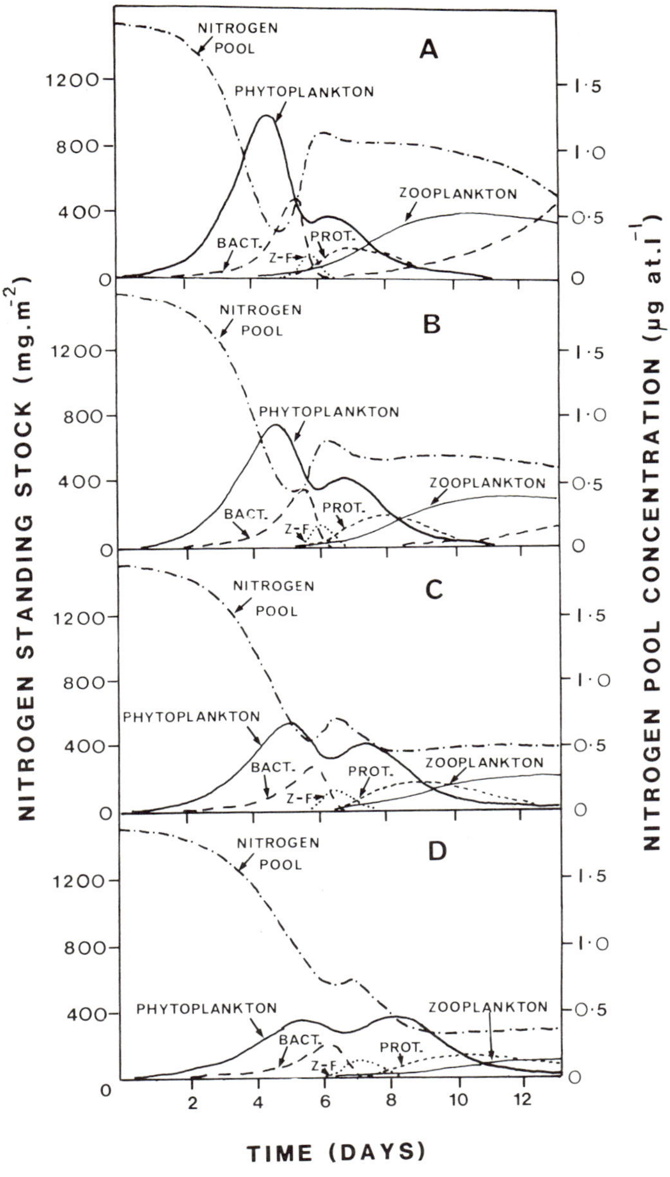

Fig. 5. Simulation results showing the development and decay of a phytoplankton bloom and the changes in the standing stocks of the associated heterotrophic compartments with increasing daily rates of sedimentation out of the mixed layer. (A) no sedimentation, (B) 10% sedimentation, (C) 20% sedimentation, (D) 30% sedimentation. Axes give nitrogen standing stocks (mg.N.m^{-2} integrated over 60 m depth) and the nitrogen pool concentration (µg.at.l^{-1}) (After Moloney et al., 1985).

corresponds to the decline of the phytoplankton bloom, as has been observed in nature (Parsons et al., 1977 ; Sorokin, 1977). The phytoplankton bloom persists for about 11 days from start to finish, before being grazed down by zooplankton and large protozoa. Blooms of comparable duration have been observed in the North Sea (Weichart, 1980 ; Gieskes and Kraay, 1980) and in the Celtic Sea (Fasham et al., 1983). The recovery of the phytoplankton bloom at day 6 when the microbial standing stocks are high and the nitrogen pool increasing suggests that microbes and indirectly responsible. At this stage, phytoplankton growth is attributed to "regenerated production" (Dugdale and Goering, 1967; Yentsch et al., 1977). The abiotic nitrogen pool remained artificially high during the simulation, which may be attributed to the closed model, with no losses of nitrogen from the mixed layer by migration, mixing, diffusion or sedimentation.

To investigate the vertical losses of nitrogen from the mixed layer by sedimentation, we simulated daily losses of 10 %, 20 % and 30 % of the phytoplankton standing stock and zooplankton faeces. Figs 5 B-D show the results of increasing rates of sedimentation. Several trends emerge. The nitrogen pool declines with increasing rates of sedimentation and all the standing stock peaks are lower. It is also notable that with increasing rates of sedimentation the "new production" phytoplankton peak is smaller and the second "regenerated production" peak assumes greater relative importance in the whole bloom although it is no higher than when there is no sedimentation.

Finally, Fig. 5 shows that all the standing stock peaks tend to be flatter and shifted to the right with increasing sedimentation, because the phytoplankton biomass is eroded continuously by sedimentation resulting in a delay in development of the bloom. The smaller biomass of phytoplankton in the upper mixed layer results in less demand on the nitrogen pool, allowing a smaller biomass to grow at a faster mass specific rate. Thus after 13 days at a sedimentation rate of 30 % per day, cumulative phytoplankton production was 77 % of that when no sedimentation was modelled. Vertical transport of phytoplankton and zooplankton faeces out of the mixed layer thus appears to maintain the phytoplankton in a state of exponential growth, with reduced biomass but faster turnover rate (P/B).

CONCLUSION

Both of the models presented here are based on the flow of nitrogen through ecosystems : in one case a nearshore kelp-bed system and in the other a general offshore plankton community. The klep bed model was developed to explore the hypothesis that nitrogen flow is affected by horizontal water transport in shallow water marine systems ; here wave action or mixing associated with horizontal transport are likely to retain nitrogen in the photic zone and the benthic community is of fixed location so that boundaries of the system can be easily defined. In pelagic systems, on the other hand, the community tends to move horizontally with water in the mixed layer, and vertical transport into and out of the mixed layer is an important feature of the system dynamics.

The model of nitrogen flow through a kelp bed shows that water transport has a strong influence on nitrogen flow through benthic communities which are dominated by filter-feeders. The magnitude of water transport may be such that budgets or models that are based on the assumption that the system is closed may be misleading. Water transport has an effect on the sensitivity of the model to biological variables, such as feeding efficiencies and assimilation efficiencies. When modelled as a closed system, the kelp bed community is shown to depend largely upon detritus and the microbial community associated with the breakdown of macrophyte debris and faeces. This is probably analagous to the situation in closed bays and estuaries with little net outflow. Under downwelling conditions in the kelp-bed, phytoplankton is transported into the system in a manner analagous to that on exposed intertidal rocky shores (Field, 1983). Under these conditions, the benthos feeds mainly on phytoplankton and the rate of supply of phytoplankton from the large offshore reservoir is as important as the rate of phytoplankton primary production.

If one models the flow of nitrogen under upwelling conditions when cool, clear water is brought into the shallow water from below the photic zone, there is net loss of organic nitrogen from the system. A diminished food supply of detritus is thus available to the benthic community, so that a smaller standing stock would be maintained if such upwelling was maintained continuously. This may be unique to nearshore upwelling systems, although there are similarities with estuaries subjected to flushing.

In deeper waters phytoplankton blooms have been modelled on a seasonal scale (e.g. Riley et al., 1949 ; Steele, 1974 ; Pace et al., 1984 ; Fasham, 1985), whereas we have presented a model with a short time-scale to explore the build-up and decline of phytoplankton and microbial populations after an upwelling or mixing event. The simulation of changes in a plankton community on a daily time scale shows that it is likely that very different biomasses of phytoplankton, bacteria, flagellates, larger protozoa, and micro/mesozooplankton do exist at different stages in the development of a bloom after nitrate enrichment. The model depicts a continuum of successive stages in which phytoplankton, microbes, and zooplankton each form major components of the biomass at different times, offering an explanation for the conflicting results obtained from samples taken at sea. Most observations have been vertical profiles at one or more stations, giving an instant "snapshot" of the relative quantities of different components of the system. The large peaks of bacterial standing stocks simulated from about day 6 in some simulations on the model agree with values recorded by Sorokin (1975, 1981) and Sieburth et al. (1978). The simulated bacterial populations eventually decline due to grazing by zooflagellates (see Fenchel, 1982 ; Azam et al., 1983). When sedimentation of faeces and phytoplankton is simulated in the model, micro- and meso-zooplankton assume greater importance from about day 9 onwards. This may correspond to the quasi-steady state conditions which occur in temperate stratified waters during the late summer. Thus different components dominate the plankton community at different successional stages of the model plankton bloom in the upper layers of the open sea water column, and the dominance of the components in the succession is influenced by the amount of material which sediments through the pycnocline and out of the surface layers.

It is hoped that this contribution will add emphasis to the importance of water transport as a factor influencing the structure and functioning of nearshore marine ecosystems, and help resolve the controversy over the significance of micro-organisms in the plankton community, which varies according to the time elapsed after an enrichment event and the amount of material sedimenting out of the surface layer. Both of these are problems which have been difficult to resolve by direct observation and measurement in the sea. Simple simulation models, based on hypotheses about the processes involved, produce output which appear to be consistent with observations, thus helping to further understanding.

ACKNOWLEDGMENTS

We thank M.I. Lucas, M.O. Bergh, R.C. Newell, F.V. Wulff and S.J. Painting for help and discussion in developing the models, and S. Tolosana and D. Gianakouras for help in preparing the manuscript. The work was supported by the South Africa National Committee for Oceanographic Research as part of the Systems Analysis Project of the Benguela Ecology Programme.

REFERENCES

Azam, F., Fenchel, T., Field, J.G., Gray, J.S., Meyer-Reil, L.A. and Thingstad, F., 1983. The ecological role of water-column microbes in the sea. Mar. Ecol. Prog. Ser., 10: 257-263.

Bienfang, P.K. and Harrison, P.J., 1984. Co-variation of sinking rate and cell quota among nutrient replete marine phytoplankton. Mar. Ecol. Prog. Ser., 14: 297-300.

Billett, D.S.M., Lampitt, R.S., Rice, A.L. and Mantoura, R.F.C., 1983. Seasonal sedimentation of phytoplankton to the deep-sea benthos. Nature, 302: 520-522.

Brown, P.C., 1984. Primary production at two contrasting nearshore sites in the southern Benguela upwelling region, 1977-1979. S. Afr. J. mar. Sci., 2: 205-215.

Dugdale, R.C. and Goering, J.J., 1967. Uptake of new and regenerated forms of nitrogen in primary productivity. Limnol. Oceanogr., 12: 196-206.

Eppley, R.W., Rogers, J.N. and McCarthy, J.J., 1969. Half-saturation constants for uptake of nitrate and ammonium by marine phytoplankton. Limnol. Oceanogr., 14: 912-920.

Fasham, M.J.R., 1985. A flow analysis of materials in the marine euphotic zone. In: R.E. Ulanowicz and T. Platt (eds), Ecosystem Theory for Biological Oceanography. Can. Bull. Fish. Aquat. Sci., 213 : 139-162.

Fasham, M.J.R., Holligan, P.M. and Pugh, P.R., 1983. The spatial and temporal development of the spring phytoplankton bloom in the Celtic Sea, April 1979. Prog. Oceanog., 12: 87-145.

Fenchel, T., 1982. Ecology of heterotrophic microflagellates IV. Quantitative occurrence and importance as consumers of bacteria. Mar. Ecol. Prog. Ser., 9: 35-42.

Field, J.G., 1983. Flow patterns of energy and matter. In: O. Kinne (ed.), Marine Ecology Vol. 5, Part. 2, Wiley, Chichester, U.K., pp. 758-794.

Field, J.G., Griffiths, C.L., Griffiths, R.J., Jarman, N., Zoutendyk, P., Velimirov, B. and Bowes, A., 1980 a. Variation in structure and biomass of kelp communities along the southwest Cape coast. Trans. R. Soc. Afr., 44: 145-203.

Field, J.G., Griffiths, C.L., Linley, E.A.S., Carter, R.A. and Zoutendyk, P., 1980 b. Upwelling in a nearshore marine ecosystem and its biological implications. Estuar. coast. mar. Sci., 11: 133-150.

Field, J.G., Jarman, N.G., Dieckmann, G.S., Griffiths, C.L., Velimirov, B. and Zoutendyk, P., 1977. Sun, waves, seaweed and lobsters : the dynamics of a west coast kelp-bed. S. Afr. J. Sci., 73: 7-10.

Gieskes, W.W.C. and Kraay, G.W., 1980. Primary productivity and phytoplankton pigment measurements in the northern North Sea during FLEX 76. "Meteor" Forschungsergebnisse, Reihe A., n° 22: 105-112.

Griffiths, C.L. and King, J.A., 1979 a. Some relationships between size, food availability and energy balance in the ribbed mussel Aulacomya ater. Mar. Biol., 51: 141-149.

Griffiths, C.L. and King, J.A., 1979 b. Energy expended on growth and gonad output in the ribbed mussel Aulacomya ater. Mar. Biol., 53: 217-222.

Harrison, W.G., Eppley, R.W. and Renger, E.H., 1977. Phytoplankton nitrogen metabolism, nitrogen budgets, and observations on copper toxicity : controlled ecosystem pollution experiment. Bull. Mar. Sci., 27: 44-57.

Holligan, P.M., Harris, R.P., Newell, R.C., Harbour, D.S., Head, R.N., Linley, E.A.S., Lucas, M.I., Tranter, P.R.G. and Weekley, C.M., 1984 a. Vertical distribution and partitioning of organic carbon in mixed frontal and stratified waters of the English Channel. Mar. Ecol. Prog. Ser., 14: 111-127.

Holligan, P.M., Williams, P.J. le B., Purdine, D. and Harris, R.P., 1984 b. Photosynthesis, respiration and nitrogen supply of plankton populations in stratified frontal and tidally mixed shelf waters. Mar. Ecol. Prog. Ser., 17: 201-213.

Honjo, S., 1978. Sedimentation of materials in the Sargasso Sea at a 5,367 m deep station. Deep-Sea Res., 36: 469-492.

Jansson, B.O., Wilmot, W. and Wulff, F., 1982. Coupling the sub-systems- the Baltic Sea as a case study. In: M.J.R. Fasham (ed.), Flows of energy and materials in marine ecosystems: theory and practice. Plenum Press, New York, pp. 549-595.

Joint, I.R. and Morris, R.J., 1982. The role of bacteria in the turnover of organic matter in the sea. Oceanogr. Mar. Biol. Ann. Rev., 20: 65-118.

Joiris, C., Billen, G., Lancelot, C., Dard, M.H., Mommaerts, J.P., Bertels, A., Bosicart, M., Nijs, J. and Hecq, J.H., 1982. A budget of carbon cycling in the Belgian coastal zone; relative roles of zooplankton, bacterioplankton and benthos in the utilization of primary production. Neth. J. Sea Res., 16: 260-275.

Koop, K., Newell, R.C. and Lucas, M.I., 1982. Microbial regeneration of nutrients from the decomposition of macrophyte debris on the shore. Mar. Ecol. Prog. Ser., 9: 91-96.

Linley, E.A.S. and Newell, R.C., 1985. Estimates of bacterial growth yields based on plant detritus. Bull. Mar. Sci., (in press).

Longhurst, A.R., 1983. Benthic-pelagic coupling and export of organic carbon from a tropical Atlantic continental shelf - Sierra Leone. Estuar. coastl. Shelf Sci., 17: 261-285.

Lugo, A.E., Sell, M. and Snedaker, S.C., 1976. Mangrove ecosystem analysis. In: B.C. Patten (ed.), Systems Analysis and Simulation in Ecology. Academic Press, New York, pp. 114-146.

McIsaac, J.J. and Dugdale, R.C., 1972. Interactions of light and organic nitrogen in controlling nitrogen uptake in the sea. Deep-Sea Res., 19: 209-232.

Mann, K.H., 1985. The role of detritus at the land-sea boundary. This volume.

Miller, R.J., Mann, K.H. and Scarratt, D.J., 1971. Production potential of a seaweed-lobster community in eastern Canada. J. Fish. Res. Bd Can., 28: 1733-1738.

Moloney, C.L., Bergh, M.O., Field, J.G. and Newell, R.C., 1985. The effect of sedimentation and microbial nitrogen regeneration in a plankton community : a simulation investigation. J. Plank. Res. (submitted).

Monod, J., 1949. The growth of bacterial cultures. Ann. Rev. Microbiol., 3: 371-394.

Newell, R.C., 1965. The role of detritus in the nutrition of two marine deposit feeders, the prosobranch Hydrobia ulvae and the bivale Macoma balthica. Proc. zool. Soc., 144: 25-45.

Newell, R.C. and Field, J.G., 1983 a. The contribution of bacteria and detritus to carbon and nitrogen flow in a benthic community. Mar. Biol. Lett., 4: 23-36.

Newell, R.C. and Field, J.G., 1983 b. Relative flux of carbon and nitrogen in a kelp-dominated system. Mar. Biol. Lett., 4: 249-257.

Newell, R.C., Field, J.G. and Griffiths, C.L., 1982. Energy balance and significance of microorganisms in a kelp bed community. Mar. Ecol. Prog. Ser., 8: 103-113.

Newell, R.C. and Linley, E.A.S., 1984. The significance of microheterotrophs in carbon and nitrogen flow through plankton communities in the western English Channel. Mar. Ecol. Prog. Ser., 16: 105-119.

Newell, R.C. and Lucas, M.I., 1981. The quantitative significance of dissolved and particulate organic matter released during fragmentation of kelp in coastal waters. Kieler Meeresforsch., 5: 356-369.

Nixon, S.W., 1985. The retention of nutrients (C, N, P), heavy metals (Mn, Ce, Pb, Cu), and petroleum hydrocarbons in Narrangansett Bay. This volume.

Pace, M.L., Glasser, J.E. and Pomeroy, L.R., 1984. A simulation analysis of continental shelf food webs. Mar. Biol., 82: 47-63.

Parsons, T.R., Takahashi, M. and Hargrave, B., 1977. Biological Oceanographic Processes. Pergamon Press, Oxford, 332 pp.

Pomeroy, L.R., 1979. Secondary production mechanisms of continental shelf communities. In: R.J. Livingstone (ed.). Ecological processes in coastal and marine ecosystems, Plenum Press, New York, pp. 163-186.

Pomeroy, L.R. and Wiegert, R.G., 1981. The ecology of a salt marsh. Springer-Verlag, New York.

Rice, D.L., 1982. The detritus nitrogen problem : new observations and perspectives from organic chemistry. Mar. Ecol. Prog. Ser., 9: 153-162.

Riley, G.A., Stommel, H. and Bumpus, D.A., 1949. Quantitative ecology of the plankton of the western North Atlantic. Bull. Bingham Oceanogr. Coll., 12: 1-169.

Ryther, J.H., 1969. Photosynthetis and fish production in the sea. Science, 166: 72-76.

Shafir, A. and Filed, J.G., 1980. Importance of a small carnivorous isopod in energy transfer. Mar. Ecol. Prog. Ser., 15: 109-116.

Shannon, L.V. and Field, J.G., 1985. The relation between phytoplankton production and pelagic fish in the South Benguela region. Mar. Ecol. Prog. Ser. (in press).

Sieburth, J.M., Smetacek, V. and Lenz, J., 1978. Pelagic ecosystem structure : heterotrophic compartments of the plankton and their relationships to plankton size fractions. Limnol. Oceanogr., 23: 1256-1263.

Smayda, T.J., 1970. The suspension and sinking of phytoplankton in the sea. Oceanogr. Mar. Biol. Ann. Rev., 8: 353-414.

Smetacek, V., 1984. The supply of food to the benthos. In: M.J.R. Fasham (ed.), Flows of energy and materials in marine ecosystems. Plenum Press, New York, pp. 517-547.

Sorokin, Y.I., 1975. Heterotrophic microplankton as a component of marine ecosystems. J. Obschei biologii. M., 36: 716-730.

Sorokin, Y.I., 1977. The heterotrophic phase of phytoplankton succession in the Japan Sea. Mar. Biol., 41: 107-117.

Sorokin, Y.I., 1981. Microheterotrophic organisms in marine ecosystems. In: A.R. Longhurst (ed.), Analysis of marine ecosystems. Academic Press, New York, pp. 293-342.

Steele, J.H., 1974. The structure of marine ecosystems. Blackwell, Oxford, 128 pp.

Stuart, V., 1982. Absorbed ration, respiratory costs and resultant scope for growth in the mussel Aulacomya ater (Molina) fed on adiet of kelp detritus of different ages. Mar. Biol. Lett., 3: 289-306.

Stuart, V., Field, J.G. and Newell, R.C., 1982 a. Evidence for absorption of kelp detritus by the ribbed mussel, Aulacomya ater using a new ^{57}Cr-labelled microsphere technique. Mar. Ecol. Prog. Ser., 9: 263-271.

Stuart, V. and Klumpp, D.W., 1984. Evidence for food-resource partitioning by kelp-bed filter feeders. Mar. Ecol. Prog. Ser., 16: 27-37.

Stuart, V., Newell, R.C. and Lucas, M.I., 1982 b. Conversion of kelp debris and faecal material from the mussel Aulacomya ater by marine organisms. Mar. Ecol. Prog. Ser., 7: 47-57.

Velimirov, B., Field, J.G., Griffiths, C.L. and Zoutendyk, P., 1977. The ecology of kelp bed communities in the Benguela upwelling system. Analysis of biomass and spatial distribution. Helgolander wiss. Meeresunters, 30 : 495-518.

Walsh, J.J., Rowe, G.T., Iverson, R.L. and McRoy, C.P., 1981. Biological export of shelf carbon is a sink of the global CO_2 cycle. Nature, Lond., 291: 196-201.

Warwick, R.M., Joint, I.R. and Radford, P.J., 1979. Secondary production of the benthos in an estuarine environment. In: R.L. Jeffries and A.J. Davies (eds.), Ecological Processes in Coastal Enviroments. Blackwell, Oxford, pp. 429-450.

Weichart, G., 1980. Chemical changes and primary production in the Fladen Ground area (North Sea) during the first phase of a spring phytoplankton bloom. "Meteor" Forschungsergebnisse, Reihe A., n° 22: 79-86.

Wickens, P.A. and Field, J.G., 1985. The effect of water transport on nitrogen flow through a kelp bed community. S. Afr. J. mar. Sci. (submitted).

Wulff, F.V. and Field, J.G., 1983. Importance of different trophic pathways in a nearshore benthic community under upwelling and downwelling conditions. Mar. Ecol. Prog. Ser., 12: 217-228.

Yentsch, C.M., Yentsch, C.S. and Strube, L.R., 1977. Variations in ammonium enhancement, an indication of nitrogen deficiency in New England coastal phytoplankton populations. J. Mar. Res., 35: 537.

THE RETENTION OF NUTRIENTS (C, N, P), HEAVY METALS (Mn, Cd, Pb, Cu), AND PETROLEUM HYDROCARBONS IN NARRAGANSETT BAY

S.W. NIXON, C.D. HUNT and B.L. NOWICKI

Graduate School of Oceanography, University of Rhode Island, Kingston, RI 02881 (U.S.A.)

ABSTRACT

Nixon, S.W., Hunt, C.D. and Nowicki, B.L., 1986. The retention of nutrients (C, N, P), heavy metals (Mn, Cd, Pb, Cu), and petroleum hydrocarbons in Narragansett Bay. In: P. Lasserre and J.M. Martin (eds), Biogeochemical Processes at the Land-Sea Boundary. Elsevier, Amsterdam.

Various workers have assembled relatively complete assessments of the annual input of the major biological nutrients (C, N, P), certain heavy metals (Mn, Cd, Pb, Cu), and petroleum hydrocarbons to Narragansett Bay. Other studies have developed inventories of the amounts of these materials in the sediments of the Bay. We have brought these data together with information on sediment accumulation rates in the Bay to determine the degree to which this one estuary serves as a sink for different types of materials in their passage between land and the coastal ocean.

It appears that Narragansett Bay retains less than 5 % of the nutrients, less than 10 % of the Mn, and perhaps 15-30 % of the Cd that is input to the system each year. The removal of Cu (70-95 %) and Pb (80-100 %) is much more effective. Somewhere between 25-65 % of the petroleum hydrocarbons entering the Bay remain in the sediments. These estimates are in agreement with the behavior of the different materials in sediment-water flux measurements and in experiments using the large MERL mesocosms.

INTRODUCTION

Because of their position between the upland drainage and the sea, it is reasonable to suppose that estuaries have always been subjected to appreciable loadings of nutrients and other weathering products from their watersheds. However, various other features of estuaries (fisheries, commerce, etc.) have also long made them attractive for human settlement, with the result that a far larger anthropogenic input of nutrients, heavy metals and other pollutants has often been superimposed on an already rich background.

The anthropogenic input of nutrients probably accelerated greatly with the development of urban water systems, indoor plumbing, and the "water closet". In his history of pollution in the tidal Thames, Wood (1982) noted that after 1815 it was "mandatory to connect cesspools to the sewers", and that indoor flush toilets were "widely used in London after 1830". In the United States, about one-third of urban households had running water and flush toilets by 1830 (Melosi, 1981) and between 1906-1915, Baltimore, Maryland became, "the last of the great American cities to build a modern system for sewage" (Capper, Power and Shivers, 1983). While urban areas always drained their open sewers and road runoff directly to adjacent waters, the new city sewer systems became a very large point source of nutrients and other pollutants that had formerly been put on or under the ground.

It is also likely that nutrient loadings to estuaries have been greatly increased by the use of commercial fertilizers in agriculture. In the United States, the use of inorganic fertilizer has increased approximately exponentially since 1860, with the result that over 500 times more nitrogen was applied to the fields in 1980 than in 1880 (Fig. 1).

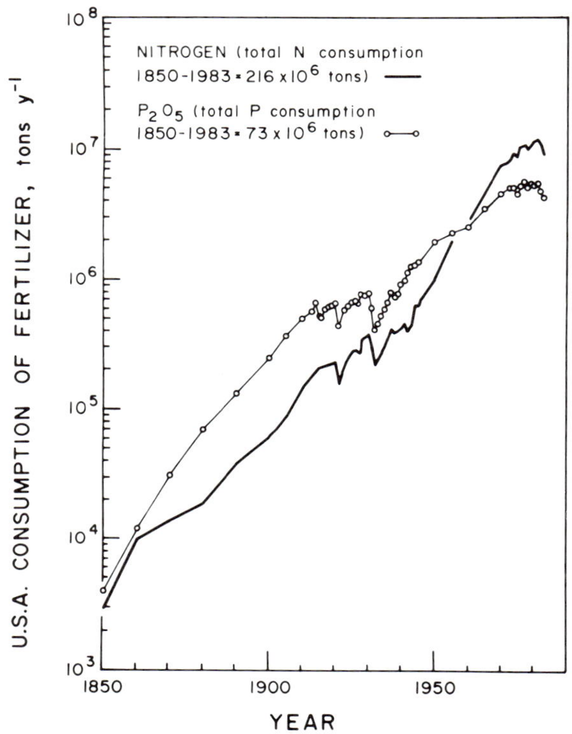

Fig. 1. Historical use of nitrogen and phosphorus fertilizer in the United States. Data from the Fertilizer Institute, Washington, DC.

The increased input of nutrients and other pollutants to rivers and estuaries may have been exacerbated by the destruction of large areas of fresh and salt water wetlands, often for agriculture (Figs 2 and 3). For example, in the United States, in 1950 there were over 100,000,000 acres (over 40 x 10^6 hectares) of wetlands being drained (Shaw and Fredine, 1956). The impact of wetlands on water quality is difficult to quantify (Nixon, 1980), but the few annual mass balance studies that are available indicate that these environments serve as sinks (though of widely varying strength) for nutrients (Fig. 4) and a number of heavy metals (see reviews by Giblin, 1982, Nixon and Lee, in press). As the fresh water swamps and marshes lining streams and rivers have been converted to urban and agricultural land, potential sites for removing sediments and nutrients before they could reach the estuary have been converted into strong sources of additional input.

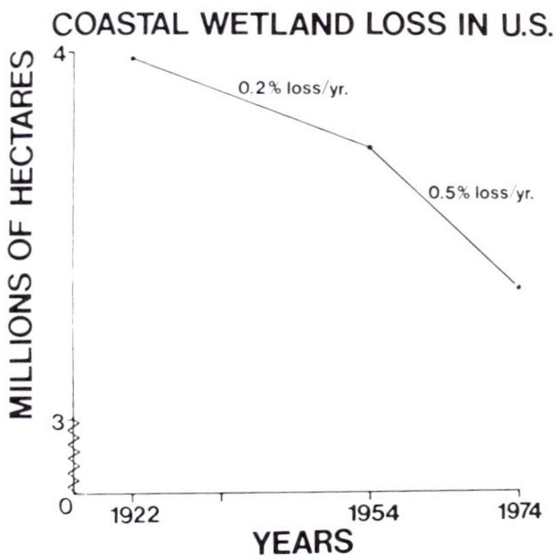

Fig. 2. Rates of coastal wetland loss in the United States. From Gosselink and Baumann (1980).

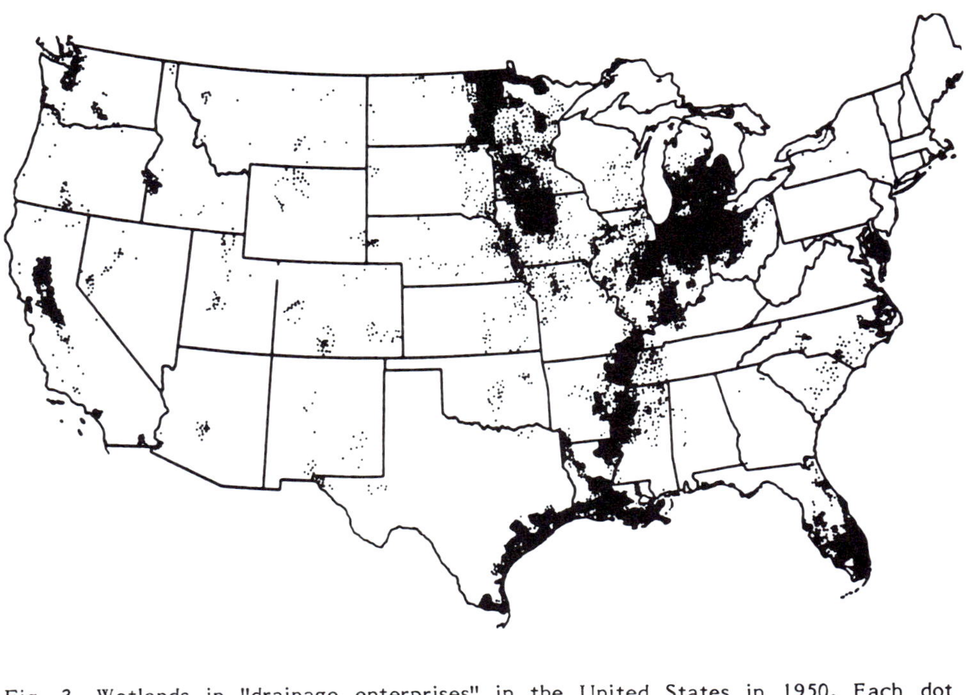

Fig. 3. Wetlands in "drainage enterprises" in the United States in 1950. Each dot represents 10,000 acres (over 4000 hectares). From Shaw and Fredine (1956).

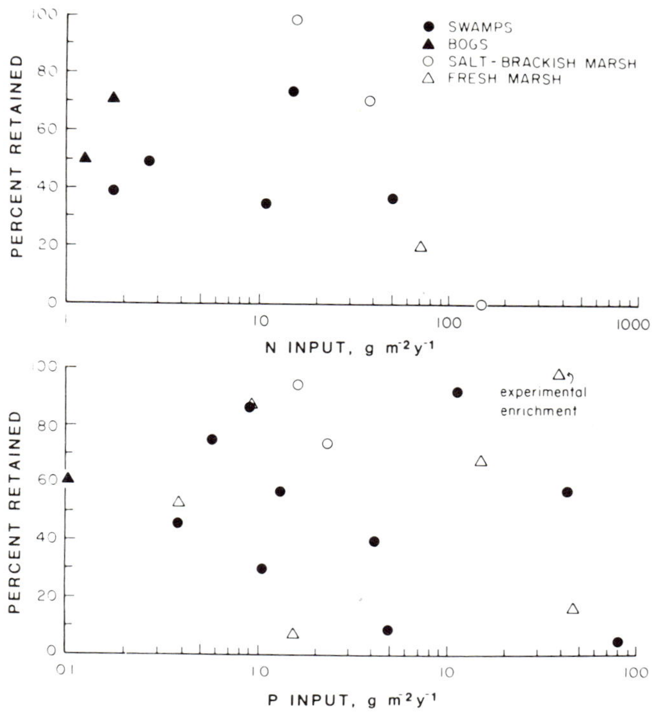

Fig. 4. The results of annual mass balance studies for a number of fresh and salt water wetland areas receiving different levels of nutrient input. From Nixon and Lee (in press).

As a result of their geographical circumstances and these various historical developments, many estuaries appear to be among the most heavily loaded ecosystems in the world in terms of the amount of nitrogen and phosphorus received per unit area (Fig. 5). Even the farm fields of the American corn belt do not receive the fertilization common to many estuaries, and the nutrient inputs to natural terrestrial systems and fresh waters appear to be orders of magnitude lower. There are fewer data available on the inputs of metals and other pollutants to estuaries, but the evidence at hand suggests that the delivery of heavy metals to estuaries may equal or exceed that deposited from the atmosphere on heavily industrialized urban areas and exceed that deposited on rural terrestrial ecosystems by orders of magnitude (Fig. 6).

Given such large inputs (even though their exact values must be subject to considerable variation and uncertainly), it is important to determine how much of the nutrients and other materials which enter an estuary pass through to the coastal ocean. Since it is extremely difficult to obtain direct measurements of the net transport of dissolved or particulate substances at the mouth of the estuary, the most promising

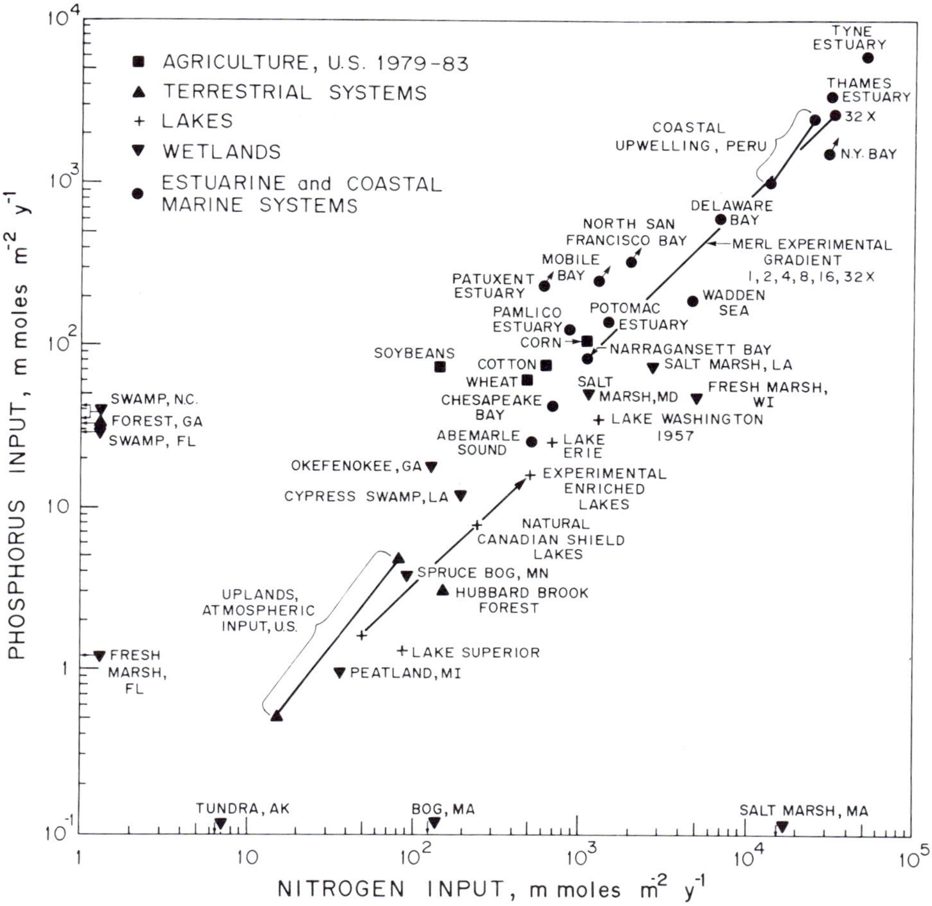

Fig. 5. Approximate present day annual rates of input of nitrogen and phosphorus per unit area of terrestrial, lake, wetland, agricultural and estuarine or coastal marine ecosystems. Systems for which the input of only one element was reported are shown. along the appropriate axis. The estuarine areas with arrows attached only include dissolved inorganic nutrients. Their rates would be increased considerably if total N and P were used. Atmospheric and wetland inputs from numerous sources summarized by Nixon and Lee (in press) ; Canadian Lakes from Schindler (1974) ; Lakes Superior and Erie from Jaworski (1981) ; Lake Washington from Edmondson (1969) ; Agriculture from Andrilenas and Serletis (1984) ; Wadden Sea from Postma and Dijkema (1982) ; other estuaries from Jaworski (1981) for total inputs and Nixon (1983) for inorganic inputs. The MERL eutrophication experiment is described in Nixon et al. (1984). The estimate for the peruvian upwelling is very approximate based on information given by Dugdale (1976) and Walsh and Dugdale (1971) assuming 6 mo y^{-1} of active transport.

approach in addressing this problem is to estimate the rates at which any material of interest is accumulating in the estuary and/or being transformed into gases which can exchange with the atmosphere. The difference between the sum of accumulation (burial in

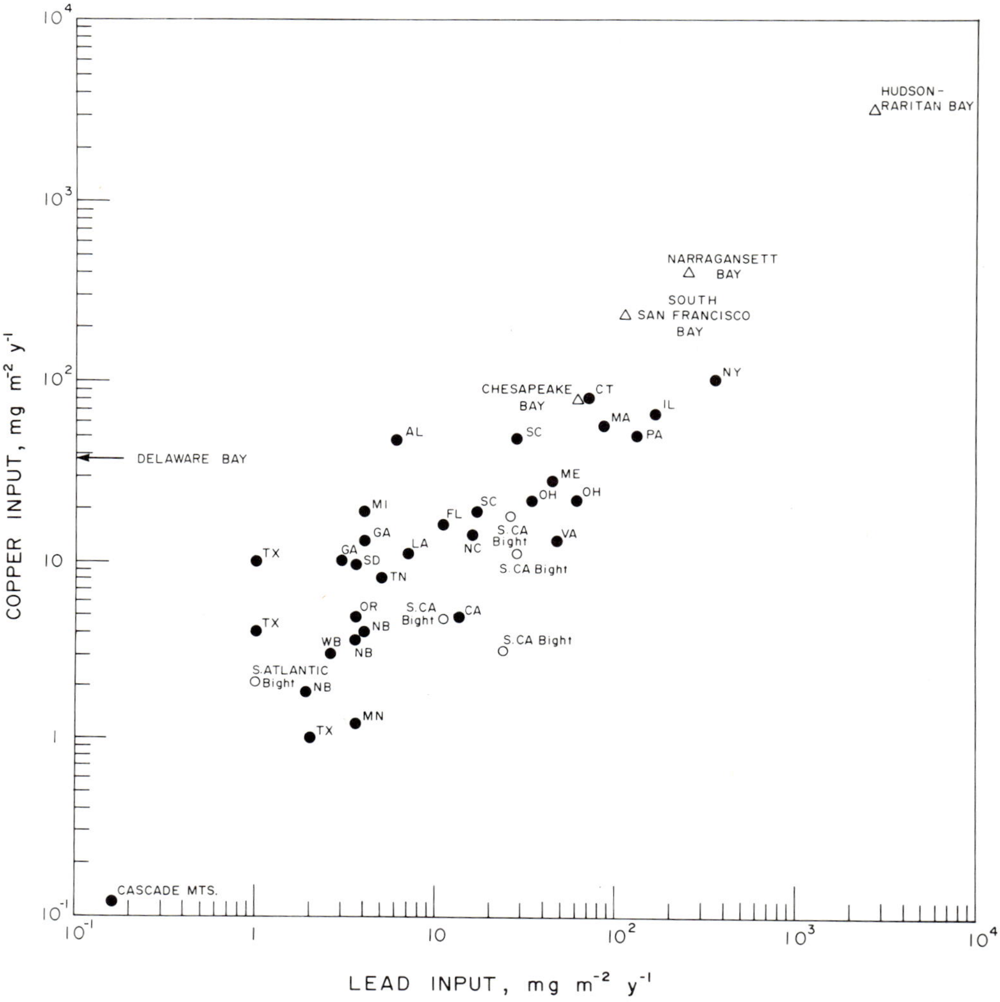

Fig. 6. Approximate present-day annual rates of input of lead and copper per unit area of various terrestrial and estuarine ecosystems in the United States. Data for upland areas from numerous sources summarized by Nixon and Lee (in press). Estuarine inputs from T.M. Church, University of Delaware (person.communication) for Delaware Bay ; Bradford and Luoma (1980) for San Francisco Bay ; Bieri et al. (1982) for Chesapeake Bay ; Mueller, Gerrish and Casey (1982) for the Hudson-Raritan estuary ; various sources noted in this paper for Narragansett Bay.

sediments) plus losses to the atmosphere and the input can then be assumed to equal the net transport to offshore waters. Removal of materials in fishery harvest must usually be a very small fraction of the mass balance, but could be relatively easily estimated in any case.

In this paper we have brought together much of the information that is presently available on certain pollutants in Narragansett Bay (Rhode Island, U.S.A.) in an attempt to develop annual mass balances for carbon, nitrogen and phosphorus as well as some heavy metals (Mn, Cd, Pb, Cu) and petroleum hydrocarbons under conditions of recent input. We cannot pretend complete knowledge of the quantities or behavior of any of these substances in the Bay, but they have all been subjected to considerable study and it seemed that it would be interesting and instructive to compare the effectiveness of this one system as a trap for a variety of different materials.

Narragansett Bay

Narragansett Bay lies in the southern New England region of the northeastern United States. The Bay proper covers about 265 km^2 (with an additional 63 km^2 in the adjacent and connected Mt. Hope Bay and Sakonnet River), and has an average depth at mean low water of about 8.4 m. A hypsographic analysis shows that 70 % of the Bay is shallower than 10 m. The tide is semidiurnal and ranges from about 1.1 m at the mouth to 1.4 m at the head of the Bay. Annual temperatures range from -0.5°C to 24°C. Fresh water input for the entire 328 km^2 area comes from a drainage basin of some 4700 km^2 and averages about 100 m^3s^{-1}. Rainfall directly on the Bay adds, on the average, another 5 m^3s^{-1}. Salinities are high and relatively constant over the annual cycle, at least in the lower Bay. Since the fresh water enters over dams, there is no well-developed mixing zone, and salinities range from about 20-24 °/₀₀ in the Providence River to 31-33 °/₀₀ at the mouth of the Bay. In general, the waters are also well-mixed, and there is only weak vertical stratification except in the upper Bay and Providence River. The residence time of the water appears to range from about 10-40 days, with a long-term mean of 25-30 days.

The Providence metropolitan area is located at the head of Narragansett Bay and (along with some other minor inputs) directly discharges into the Bay some 0.46 x 10^6 m^3 d^{-1} of primary and secondary treated sewage containing large amounts of industrial as well as domestic wastes. In addition, some 0.11 x 10^6 m^3 d^{-1} of sewage are discharged from Massachusetts to Mt. Hope Bay, but it is not known how much of this material actually enters Narragansett Bay. The rivers entering the Bay have also passed through a number of urban industrial areas and carry a significant burden of pollutants into the system.

There is relatively little suspended sediment in most of the lower Bay (4 mg $liter^{-1}$), though values of 100 mg $liter^{-1}$ may be found at times in the Providence River. Extinction coefficients range from -0.2 m^{-1} in the lower Bay to about -1m^{-1} in the Providence River. As a result of the generally clear water and abundant nutrients (NH_4^+ = 0-10 μM, $NO_{2,3}^-$ = 0-10 μM, PO_4^- = 0.2-3 μM, $Si(OH)_4^-$ = 0-25 μM), phytoplankton production is high (perhaps 300-400 gC m^{-2} y^{-1}). The bottom of the Bay is covered by silt-clay and sand sediments, and there are only occasional dense areas of macroalgae or seagrasses around the periphery of the area. Dissolved oxygen is always near saturation, even in the bottom water, except during summer in the Providence River.

More complete descriptions of these, and other features of the system can be found in Nixon and Kremer (1977), Kremer and Nixon (1978), Pilson (1985) and the numerous references these reviews contain.

MASS BALANCES

Inputs

Developing an inventory of the annual inputs of nutrients, heavy metals, and hydrocarbons to Narragansett Bay is relatively straightforward in concept, but difficult and tedious in practice. No overall, coordinated sampling program has led to the present level of information summarized in Table 1. Rather, various studies have been made at different times of each type of material. As a result, certain inputs are better described for one material than another and the quality of the estimates varies considerably. However, it would require a lengthy manuscript to deal with this problem alone, and our intent here is to accept the values given in earlier papers as representing the best estimates available. Our impression is that the information regarding inputs for this estuary is at least as good as that for most other systems.

The position of Narragansett Bay in Figures 5 and 6 shows that it is comparatively heavily loaded in terms of both nutrients and metals. There appear to be few, if any, detailed assessments of petroleum hydrocarbon inputs to other estuaries (Olsen et al., 1982), but data given by Mueller, Gerrish and Casey (1982) show that the Hudson-Raritan estuary off New York city may receive some 325 g m^{-2} y^{-1} of "oil and grease". The summary given by Olsen et al. (1982) also suggests that in the past, lower Delaware Bay may have received hydrocarbon inputs similar in magnitude to those of Narragansett Bay, but that recent inputs to the Delaware are more like 2-3 g m^{-2} y^{-1}.

Accumulation in the Bay

It is more difficult to determine the fate of material once it has entered the Bay than it is to measure its rate of delivery to the system. A common geochemical tool, the mixing diagram (Boyle et al., 1974), is difficult to use in Narragansett Bay because there is only a very weak salinity gradient along the axis of the Bay and the major sewage inputs occur in relatively high salinity water. As a result, the system does not lend itself readily to a modeling analysis such as that used by Kaul and Froelich (1984).

There are no long-term data on the concentrations of nutrients, metals, or hydrocarbons in the waters of Narragansett Bay, but the little information that is available on nutrients in the water during the past 25 years suggests that there is no clear secular change which can be seen against the background of year-to-year variation (Nixon and Lee, 1979 ; Pilson, personal communication). The sediments of the Bay serve as a repository for any materials that are removed during the passage of the water from land to sea. Fortunately, the sediments of Narragansett Bay have been analyzed for a variety of materials, though at only a modest number of locations around the Bay. The sediment analyses have been carried out to various depths at different sites and not all of the materials of interest in this paper have been measured in all of the cores that have been collected. Nevertheless, the data set in hand is large enough to make a beginning in assessing the rate at which materials are accumulating in the bottom sediments. Major sources of data include Goldberg et al. (1977) ; Hunt and Smith (1983) ; and Santschi et al. (1984) for metals, Hurtt (1978) and Hurtt and Quinn (1979) for petroleum hydrocarbons, and Nixon and Nowicki (unpublished) for C, N and P. The data for Cu, Pb and hydrocarbons from six sites in the Bay were used with measurements of radionuclides (^{234}Th, ^{210}Pb and ^{239,240}Pu) with known input functions to calculate the net accumulation of sediments, Cu, Pb and hydrocarbons in the Bay (Santschi et al., 1984). A similar modeling was later carried out by Santschi (personal communication) for Cd in the sediments. Since C, N, P and Mn do not appear to show a clear anthropogenic enrichment in the sediments (Fig. 7 ; Nixon and Pilson, 1984), we calculated their accumulation as the product of the average concentration (below 10 cm for nutrients) and the sediment deposition rate.

The uniformity of Mn with depth in the sediments of Narragansett Bay contrasts with the marked Mn enrichment of surface sediments reported from some other estuaries, including Long Island Sound (Thomson, Turekian and McCaffrey, 1975) and the St. Lawrence (Sundby, Silverberg and Chesselet, 1981). There are large fluxes of dissolved Mn from the sediments to the overlying water in Narragansett Bay (Graham et al., 1976 ; Elderfield et al., 1981 a ; Hunt, 1983) and it may be that the metabolism of the bottom in this estuary is sufficiently great that reduced conditions are maintained too close to the sediment-water interface to allow a reprecipitation and accumulation of Mn. Perhaps the water circulation near the bottom is sufficiently vigorous that very fine particles such as those found in the St. Lawrence containing reprecipitated Mn (Sundby, Silverberg and Chesselet, 1981) are dispersed throughout the water column and transported out of the system.

A striking feature of the solid phase concentration profiles shown in Figure 7 is the early enrichment of the sediments with Pb and Cu (after about 1830-1840 at this Upper Bay site), Cd (after about 1880), and petroleum (after about 1920) compared with the relatively constant levels of C, N and P below about 10 cm (about 1960). Metal inputs to this estuary began early during the American phase of the Industrial Revolution, and by 1860 Rhode Island was the most highly industrialized state in the Union (Mc Loughlin, 1978). After about 1850, G.H. Corliss and others made Providence a major center of steam engine development and manufacture, with the result that the base metal industry rose to a prominent position in the state's economy and the output of the jewerly industry increased markedly (Coleman, 1969). Since the industrial wastes from these activities were presumably discharged to Narragansett Bay, it is not surprising to find metals relatively deep in the sediments. However, the major source of nutrients discharged to the Bay, the Field's Point Sewage Treatment Plant, began construction in 1890 and operation in the early 1900's. Yet there is virtually no reflection of this enrichment in the sediments. The higher concentrations of C, N and P in the near-surface zone reflect the recent deposition of organic materials which have been mixed to about 10 cm and which are being rapidly decomposed. Other studies have shown that the sediments of the Bay are mixed by resuspension and bioturbation to depths of 10-20 cm (McCaffrey et al., 1980 ; Elderfield et al., 1981 b ; Santschi et al., 1983), and that perhaps 5-10 % of freshly deposited organic matter is carried to sub-surface layers of the sediments (Garber, 1984). The anthropogenic input of Mn is also significant (Table 1), but can not be seen in the sediment profile (Fig. 7).

Differences in the sediment profiles of these various materials are due to differences in the balance between the efficiency and speed with which they are carried to the bottom and the rate at which they are remobilized and returned to the overlying water. Phytoplankton rapidly deplete the dissolved inorganic nutrients from the water column during the winter-spring bloom and throughout the summer (Smayda, 1973 ; Furnas, Hitchcock and Smayda, 1976 ; Furnas, 1982), and studies in the MERL mesocosm tanks (5 metres deep) by Santschi et al. (1983) and Kelly (1983) have shown a rapid transport of particles from the water column to the sediments (Table 2). Similar studies using additions of radioactive and stable metals and petroleum hydrocarbons to the MERL tanks have also shown a rapid flux to the sediments of Mn, Pb and hydrocarbons, with a somewhat slower removal of Cu and a considerably less effective removal of Cd (Table 2). Thus, some of the materials which are most rapidly and effectively carried to the bottom (C, N, P, Mn) do not remain there, while others do (Pb, petroleum hydrocarbons). Copper, which appears less particle-reactive in the water column than Pb, is effectively retained in the sediments once it arrives on the bottom. Cadmium is maintained predominantly in dissolved form (Adler, Amdurer and Santschi, 1980), but a significant portion of the small fraction reaching the bottom appears to remain as a part the sedimentary record (Fig. 7).

The evidence from sediment profiles and water column removal experiments regarding the behavior of these various materials is supported by direct measurements of their return to the overlying water after remineralization and remobilization on and in the bottom sediment (Table 3). In the case of petroleum hydrocarbons, C and N, this regeneration involves the production of CO_2 and N_2 + N_2O which may be lost to the atmosphere from the Bay or carried offshore as dissolved gases. The importance of benthic nutrient regeneration in estuaries has been clear for several years (reviewed by Nixon, 1981), but it is remarkable to see that in Narragansett Bay the remobilization and return of Mn from the sediments to the overlying water is 3-4 times greater than the input, a turnover much faster than that of the major biological nutrients. The relatively weak Cd signal in the sediments is not surprising considering that its half-removal time is slow.

Deriving a number for the benthic remineralization of petroleum hydrocarbons is particularly difficult, since the various metabolic chamber incubations used to measure the net flux of O_2, CO_2, NH_4, $NO_{2,3}$, N_2O, N_2, PO_4 and metals across the sediment water interface do not tell us what substrate is being respired.

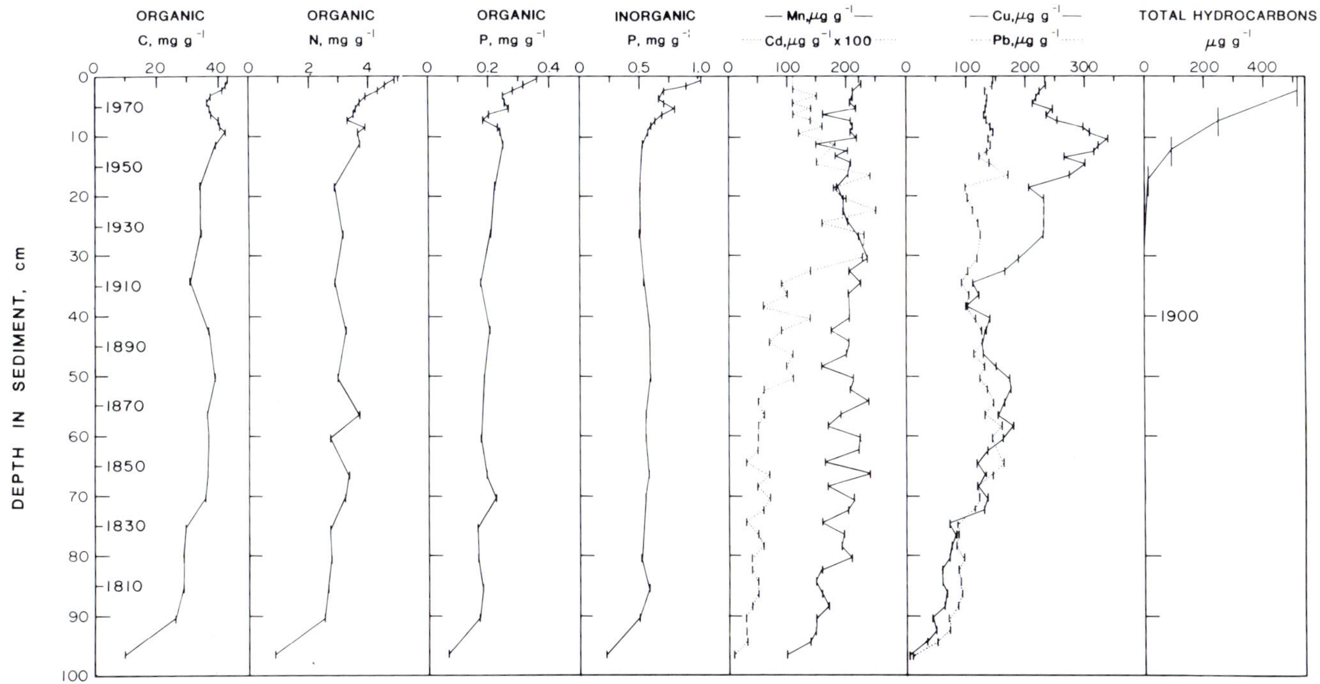

Fig. 7. Concentrations of nutrients, some heavy metals, and petroleum hydrocarbons at various depths in the sediments at Ohio Ledge in Upper Narragansett Bay. Nutrients and metals were analyzed from the same core by the authors, while hydrocarbons were measured on a separate core from a near-by site by Hurtt (1978). Geochronology of the core used for nutrients and metals is described in detail by Santschi et al. (1984).

TABLE 1

Estimates of recent inputs to Narragansett Bay (RI, USA), accumulation rates within the sediments of the Bay, and losses from the Bay to the atmosphere and offshore waters for major nutrients, various heavy metals, and total petroleum hydrocarbons.

	g m^{-2} y^{-1}			mg m^{-2} y^{-1}				g m^{-2} y^{-1}
	C	N	P	Mn	Cd	Pb	Cu	Petroleum Hydrocarbons
INPUTS[1]								
Rivers	~47	12.50	2.03	500	7.4	31.8	83.5	2.8
Sewage	~56	14.70	2.64	216	2.2	135	243	2.9
Atmosphere	~0	0.41	0.01	~35	?	~75	~70	0.003
Fixation	310	0.01	0	0	0	0	0	0.04 (spills)
Total	413	27.62	4.68	751	9.6	241.8	396.5	5.743
ACCUMULATION IN SEDIMENTS[2]	6	0.46	0.12	53	1.5-2.8	193-340	273-379	1.5-3.8
OUTPUTS								
to Atmosphere[3]	<339[a]	7.2[b]	0	0	~0	~0	0	0.9-2.0[c]
to Offshore[4]	> 74[a]	19.96[b]	4.56[b]	698[b]	6.8-8.1[b]	0-49[b]	18-124[b]	0-3.3[b]

1 : C from Furnas et al. (1976) for phytoplankton production and from the organic P reported by Nixon (1981) for rivers and sewage using a C:P ratio of 106:1 by atoms. Total N and P inputs from Nixon (1981). Sewage includes urban runoff. Metals from rivers and sewage measured by Hunt (unpublished report). Cu in rivers also includes data for small rivers and smaller sewage treatment plants reported by Hoffman and Quinn (1984). Metals from atmosphere estimated from data for nearby areas summarized by Nixon and Lee (in press). Hydrocarbons from Hoffman and Quinn (1984), and E. Hoffman (pers. comm. as reported in Santschi et al. in press) include Mt Hope Bay inputs and area.

2 : C, N and P from Nixon and Pilson (1984), Pb and Cu from Santschi et al. (in press), Mn and Cd from Santschi (pers. comm.).

3 : (a) Calculated by difference ; (b) denitrification from Seitzinger et al. (1984) ; (c) sum of volatilization and decomposition extrapolated as 47 % of input from a long-term oil addition experiment using MERL microcosms (Gearing et al., 1980).

4 : (a) From Nixon and Pilson (1984), does not include DOC or allocthonous POC ; if allocthonous DOC is carried conservatively through the system as found by Cifuentes (1982) in the Delaware estuary, then export to offshore would be < 100 g C m^{-2} y^{-1} and respiration > 303 g C m^{-2} y^{-1} ; (b) calculated by difference.

TABLE 2.

Half-removal times (days) for the flux of particles, various metals and petroleum hydrocarbons from the water column (5 m deep) to the sediments in the MERL microcosms.

	Winter	Summer
Natural Particles[1]	2	0.1
Mn[2]		1
Cd[3]	60-100	10-50
Cu[2]	40	4-10
Pb[3]	10-20	2-3
Petroleum Hydrocarbons[4]		
F_1 fraction[a]		1.9
F_2 fraction [b]		1.9

[1]Santschi et al. (1983) using the time series particle concentration and sediment trap data ; [2]C. Hunt, unpublished data from a stable metal spike addition ; [3]Santschi, Adler and Amdurer (1983) from radioactive trace metal spikes. Helf-removal times for ^{210}Pb in Narragansett Bay ranged from 2.6-5 days in summer and 22 days in winter (Santschi et al., 1983) ; [4]Gearing and Gearing (1983) from additions of No. 2 fuel oil : (a) saturated hydrocarbons and olefins with one or two double bonds--range for 9 compounds = 1.1-2 days ; (b) aromatic hydrocarbons--range for 6 compounds = 0.7-2.5 days.

The estimate given in Table 3 is based on the loss of hydrocarbons from concentration measurements in the sediments of the MERL mesocosms after a long-term experiment involving the addition of NO. 2 fuel oil (Gearing et al., 1980). The nature of the hydrocarbon materials reaching Narragansett Bay sediments is certainly more complex, and the MERL number is probably an overestimate of the biological breakdown of these compounds. Nevertheless, the widespread occurrence of petroleum as an estuarine pollutant, the extensive data on petroleum hydrocarbons in Narragansett Bay, and the unique MERL oil experiments make it compelling to assemble the budget as best we can.

Additional evidence for the almost complete regeneration of the organic matter reaching the bottom of Narragansett Bay comes from several other experiments. In one case, different amounts of seston filtered from Bay water were added to intact cores of sediment maintained in the laboratory (Kelly and Nixon, 1984). Even though the cores were maintained at only 15°C (summer bottom temperatures reach 20°), 24-30 % of the organic N and 11-20 % of the organic C added to the sediments had been remineralized within one month. The release of inorganic P was even more rapid. In a related experiment using ^{15}N labelled phytoplankton detritus, Garber (1984) concluded that, "Net rates of N-remineralization based on the conversion of labelled PON to NH_4^+ suggest that essentially all but a few percent of detrital-N reaching the sediment surface could be recycled in less than one year". Studies of the decomposition of freshly deposited organic matter from a summer phytoplankton bloom in a MERL tank also showed rapid initial remineralization rates of 10.5 %, 7.2 % and 3.1 % day^{-1} for C, N and P respectively

TABLE 3

Importance of benthic regeneration, release, or decomposition, respectively, of various nutrients, heavy metals and petroleum hydrocarbons in Narragansett Bay.

	Benthic Regeneration, Release, or Decomposition	
	Sediment-Water Flux	% of Input
	$g\ m^{-2}\ y^{-1}$	
C[1]	139.0	28
N[1]	12.4	45
N + denitrification[2]	19.6	71
P[3]	3.7	79
	$mg\ m^{-2}\ y^{-1}$	
Mn[4]	1098	416
Mn[5]	2212	295
Cd[4,6]	2.5-4.5	26-47
Cu[4,6]	0-51	0-13
Pb[6]	0-4.5	0-2
Petroleum Hydrocarbons[7]		55

[1]Nixon et al. (1976) ; [2]Seitzinger et al. (1984) ; [3]Nixon et al. (1980) ; [4] Elderfield et al. (1981 a) ; [5]Hunt (1983) ; [6]Hunt and Smith (1983) ; [7]Gearing et al. (1980) based on the loss rate in sediments during the first 100 days after inputs were terminated and applying this rate during the time of oil addition as well.

(Kelly, 1983). Even when the amount of organic matter in Narragansett Bay sediments is experimentally increased markedly, it appears that respiratory rates rise sufficiently that only a small portion remains to accumulate over an annual cycle. This result is evident in the organic enrichment of sediments described by Kelly and Nixon (1984), in the metabolic response of the sediments in the Bay to the winter-spring phytoplankton bloom (Nixon et al., 1980), an in the response of the MERL mesocosms to a long-term (2.5 yr) nutrient addition experiment (described in Nixon et al., 1984). Among the advantages of the coupled benthic-pelagic systems in the MERL tanks is the relative ease with which one can measure the total system apparent production during the day and consumption during the night based on changes in dissolved O_2 or CO_2. The results of such measurements obtained by C.A. Oviatt from replicate mesocosms receiving different levels of inorganic N, P and Si are shown in Figure 8. Only when production rose to very high levels was there a modest excess of organic matter formation over consumption, and much of this must have been flushed out of the tanks (Bay water is added to the tanks at a rate sufficient to exchange the volume in 27 days). Only a small portion of the enhanced primary production remained in the sediments. The organic-rich sediments commonly associated with polluted environments may require years of accumulation, anoxic or low oxygen conditions, and/or rapid rates of sediment deposition-all conditions which were not found in the MERL experiments or in most of Narragansett Bay.

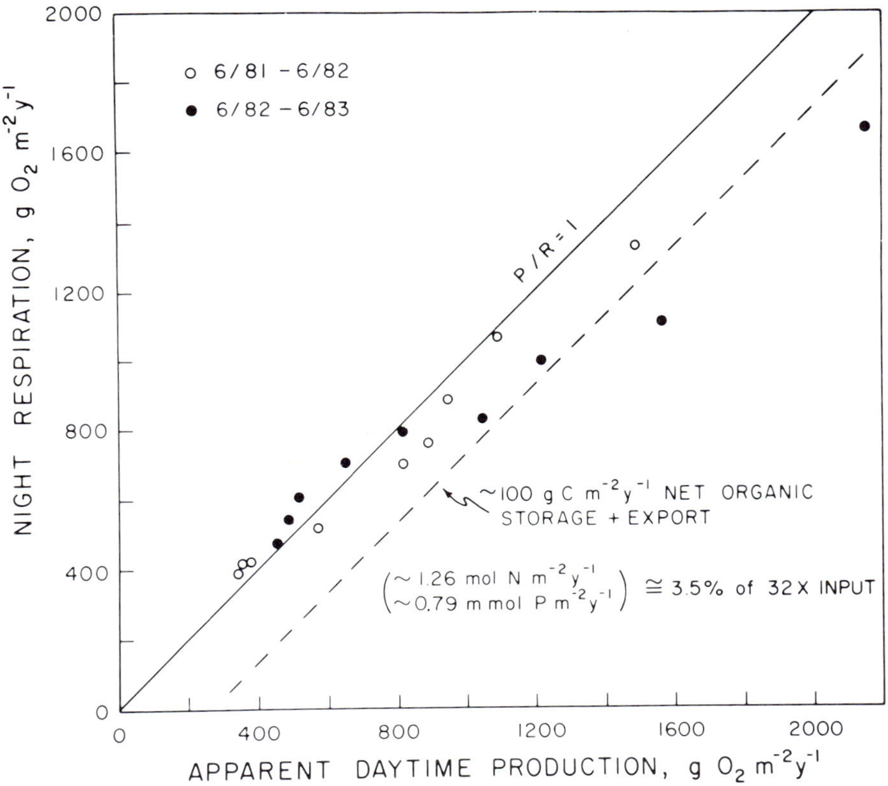

Fig. 8. The relationship between annual total system net production during the day and consumption during the dark in the MERL mesocosms during two years in which they were receiving different levels of inorganic nutrient (N, P, Si) input. Estimates are based on dawn-dusk changes in dissolved oxygen and have been corrected for air-sea gas exchange. The solid line represents a P/R ratio of 1 with no net organic accumulation in, or export from, the tanks. The broken line represents a net accumulation and/or export of 100 g C m^{-2} y^{-1} assuming a PQ of 1. Data from C. Oviatt. A description of the mesocosms and the enrichment experiment is given in Nixon et al. (1984).

Export to Offshore

A consideration of the annual mass balances summarized in Table 1 and the differing behavior of the various types of material discussed in the preceding section suggests that there are three general types of substances that differ markedly in the extent to which they will be retained within or transmitted through an estuary.

a. those that are not very effectively transported to the bottom sediments because they are not particle reactive (for example, Cd) or because they are very rapidly respired or degraded in the water (for example, low molecular weight soluble or volatile organic compounds). These materials are largely lost from the Bay.

TABLE 4

Relative efficiency (expressed as a % of the total terrestrial, anthropogenic, and atmospheric input) with which Narragansett Bay (RI, USA) removes and transmits various nutrients, heavy metals, and petroleum hydrocarbons. See Table 1 for data sources.

	C	N	P	Mn	Cd	Pb	Cu	Petroleum Hydrocarbons
% Retained	1.5	1.7	2.6	7	16-29	80-100	69-95	26-67
% Transmitted								
To Atmosphere*	82	26	0	0	0	0	0	47
To Offshore	18	72	97	93	71-84	0-20	5-31	0-27

* Some of the CO_2, N_2 and N_2O gases produced within the Bay and included in this term are carried offshore dissolved in the water rather than actually exchanged across the air-water interface in the Bay. Hydrocarbon loss includes volatilization and respiration. If allocthonous DOC is carried conservatively through the Bay, the offshore loss would rise to about 27 % of input and the flux to the atmosphere would fall to 73 %.

b. those that are carried to the bottom but which are actively remineralized and recycled (C, N, P) or resolubilized (Mn) in the sediment and released to the overlying water. These materials are also largely lost from the Bay. The export of N from the Bay is also reduced considerably because of denitrification in the sediments and the loss of N_2 and N_2O to the atmosphere (Seitzinger, Nixon and Pilson, 1984).

c. those which are carried to the bottom but which remain firmly bound to particles or otherwise trapped in the sediment (for example, Pb, Cu). These are retained and accumulate in the system.

A summary of the relative efficiency with which Narragansett Bay transmits and retains these various materials is given in Table 4 based on the absolute numbers reported in Table 1. We emphasize again that there must be a considerable (though usually undefined) uncertainly in all of these budgets. Our sense is that the input data are perhaps the most reliable part of the balance and that future efforts should be directed toward a better description of sedimentation rates and sediment characteristics in the Bay.

COMPARISON WITH OTHER ESTUARIES

In spite of the long-standing effort by limnologists, marine ecologists, and geochemists to quantify the flux of nutrients and other materials in rivers (Clarke, 1916 ; Livingstone, 1963 ; Meybeck, 1979 ; symposium volumes edited by Martin, Burton and Eisma, 1981 ; Likens et al., 1981 ; Degens, 1982), there have been remarkably few reports of the mass balances in estuaries that are needed to link riverine outputs and ocean inputs. As Bewers and Yeats (1981) concluded in their review of the behavior of trace metals during estuarine mixing, "Several global budgets involving estimates of the riverine fluxes of metals to the ocean have been constructed... However, these budgets and flux calculations

TABLE 5

Sediment input and deposition in various estuaries.

Estuary	g m^{-2} y^{-1} Input	Deposition
Long Island Sound, NY-CT[1]	157	
Narragansett Bay, RI[2]		250
Chesapeake Bay, MD-VA[3]	262	
Wadden Sea[4]	788	375
Delaware Bay[5]	1,400	
San Francisco Bay, CA[6]		3,500
Knight Inlet, B.C.[7]	19,250	
Atchafalaya Bay, LA[8]	9,680,000-19,360,000	

[1]Benninger (1978) ; [2]Santschi et al. (in press) ; [3]Bieri et al. (1982) ; [4] Postma and Dijkema (1982) ; [5]Biggs, Sharp and Howell (1984) ; [6]Conomos and Peterson (1977) ; [7]turbid Fjord draining a glacial area, Farrow, Syvitski and Tunnicliffe (1983) ; [8]shallow bay with active delta building, Van Heerden, Wells and Roberts (1983).

do not take into account the removal or additions of metals in estuaries... More emphasis upon the construction of budgets for estuaries would be justified...".

Because of this lack, it is difficult to know how common the results from Narragansett Bay may be. Before discussing some of the comparisons that can be made, however, it is worth noting a feature of the Bay that may play an important role in determining the relative strength of the estuary as a sink for materials. The best current estimate of new sediment deposition is 250 gdw m^{-2} y^{-1} averaged over the Bay bottom (Santschi et al., 1984). We have no idea how typical such a sediment input may be, but some idea of the range found in other estuaries is given in Table 5.

Nutrients

In their synthesis of many studies on the Wadden Sea, Postma and Dijkema (1982) concluded that, "... some of the phytoplankton produced in situ may be carried away to the North Sea. A small fraction will be buried in new sediment deposits. However, most of the organic matter is mineralized in the Wadden Sea...". Similarly, van Bennekom and Salomons (1981) suggest that, "... only a small part of the increase of the phosphorus load of the Rhine is trapped in coastal sediments, the rest being transported to the open North Sea". Both reports put the Wadden Sea in line with the results from Narragansett Bay. In their modeling analysis of Ochlockonee Bay, Florida (USA), Kaul and Froelich (1984) again concluded that, "all biological P-uptake is regenerated within the estuary so that virtually 100 % of the fluvial reactive-P enters the ocean". On the other hand, Loder and Glibert (1980) applied conservative mixing models to concentration data from Great Bay, New Hampshire (USA), and concluded that only 13 % of the dissolved phosphate input was exported to the coastal ocean, with the remaining fraction buried in the sediments.

The most dramatic disagreement with our results comes from the recent very large study of Chesapeake Bay carried out by the U.S. Environmental Protection Agency. Nutrient enrichment was a major concern of that study, and annual mass balances for N and P were an important part of the final result. The overall conclusion was that, "Nearly all of the materials that enter the Bay remain there ; nutrients trickle out of the Bay mouth at a very slow rate" (Smullen, Taft and MacKnis, 1982).

The authors arrived at this conclusion by comparing what appear to be very careful estimates of nutrient inputs with a very uncertain calculation of net transport across the mouth of the Bay based on a simple box model. Using another approach, however, it is possible to show that their result is almost certainly incorrect. If we accept the inputs of N (123×10^6 Kg y^{-1}), P (10×3.10^6 Kg y^{-1}), and sediment (3014×10^6 Kg y^{-1})given by Smullen, Taft and MacKnis (1982), and their conclusion that essentially all of these inputs remain in Chesapeake Bay, then the buried sediments must have an average composition of 40.9 mg N and 3.42 mg P/g dry sediment. No data are given in the report on the actual N and P concentrations in the sediments, but the steady-state concentrations in the sediments at Ohio Ledge in upper Narragansett Bay are about 3 mg N/g and 0.7 mg P/g dry sediment (Fig. 7). This is an area with silt-clay sediments where nutrient concentrations in the water are higher than in most of Chesapeake Bay. Direct evidence that Chesapeake Bay sediments are lower in N and P than the Ohio Ledge sediments is in the report by Bieri et al. (1982) that organic carbon in Chesapeake Bay sediments averages 2.2 % in Maryland and 1 % in Virginia. It only reaches 3 % in the fine sediments of the deep central Bay channel. Organic carbon in the Ohio Ledge sediments is 3-4 % (Fig. 7). If we take 2 % organic C as a working estimate for average Chesapeake Bay sediment and assume that N and P are thus 66.6 % of the value at Ohio Ledge (or 1.99 mg N and 0.466 mg P/g dry sediment), we can calculate the amount of N and P buried with the Smullen, Taft and MacKnis (1982) estimate of 3014×10^6 Kg of sediment y^{-1}. The result is 6×10^6 Kg N and 1.4×10^6 Kg P or 4.9 % of the N input and 13.6 % of the P input. If this analysis is at least nearly correct, it seems that Chesapeake Bay is behaving much like Narragansett Bay, and not at all as described by the EPA study.

Heavy Metals

In his Presidential Address before the American Geochemical Society, Turekian (1977) noted that, "The overwhelming sense is that in an estuarine system there is continuous movement of some metals in and out of solution but little is lost from the system". He illustrated his point with the study by Evans et al. (1977) of Mn in the Newport River estuary in North Carolina (USA). While the Pb and Cu budgets from Narragansett Bay support the view of a strong estuarine sink for metals, the Mn and Cd data do not. Similarly, Sundby, Silverberg and Chesselet (1981) have shown a large loss of Mn from the St. Lawrence estuary in Canada. Likewise, the conservative behvior for total Mn in Narragansett Bay (Graham et al., 1976) suggests export of most Mn entering the system.

In their studies of metals in Chesapeake Bay, however, Bieri et al. (1982) claim that more than 60 % of both the Pb and Mn input is retained in the bed sediments. In their recent studies of heavy metals in Delaware Bay (USA), Church, Tramontano and Murray (1984 and later personal communication) calculated retention of 92 % of the Mn, 37% of the Cu and 32 % of the Cd input to that estuary. However, losses from the estuary in that analysis were based on calculations of the probable flux out of the mouth of the Bay using a layered flow model. When sediment concentrations and accumulation rates were used, only small amounts of Mn and Cd appeared to be retained in the system (Church, personal communication). At this point we are not aware of any convincing evidence that clearly contradicts the findings regarding the behavior of Pb, Cu,Mn or Cd in Narragansett Bay. Unfortunately, the number of mass balances for these elements is so small that this is not a particularly reassuring claim.

The situation with regard to petroleum hydrocarbons is such that there do not appear to have been any other published attempts to develop an annual mass balance for an estuary.

The accumulation of sediments in estuaries appears to be so effective that Meade (1981) has estimated that, "Probably less than 5 percent of all river sediment discharged into the tidal waters of the Atlantic seaboard is deposited on the floor of the continental shelf or the deep sea". These sediments contain an often dramatic imprint of human activity and impact on the estuary and its watershed. But it is an imprint which results not only from the input of materials, but also from the interaction of a great variety of physical, chemical and biological processes in the estuary. We are only beginning to learn to read that record, but the early results suggest that we may learn more about what flows between rivers and the sea from the humble muds on the estuary floor than we can from the water above.

ACKNOWLEDGMENTS

We are grateful to a number of colleagues for stimulating discussions of the questions addressed in this paper and, in some cases, for permission to use their unpublished data. We particularly thank M.E. Q. Pilson,P. Santschi, C.A. Oviatt, E. Hoffman, J.G. Quinn and J.N. Gearing. Some of the chemical analyses of Narragansett Bay sediments were run by S. Northby, D. Swift and J. Hopkins. Preparation of this manuscript was made possible largely by grants from the U.S. Environmental Protection Agency and the Sea Grant Program, NOAA.

REFERENCES

Adler, D., Amdurer, M. and Santschi, P.H., 1980. Metal tracers in two marine microcosms: sensitivity to scale and configuration. In: J.P. Giesy, Jr. (ed.), Microcosms in Ecological Research. Technical Information Center, U.S. Dept. of Energy, CONF-781101, Springfield, VA, pp. 348-368.

Andrilenas, P. and Serletis, B., 1984. Inputs outlook and situation, February 1984. USDA, Econ. Res. Ser. U.S. Government Printing Office, Washington, DC.

Bennekom, A.J., van, and Salomons, W., 1981. Pathways of nutrients and organic matter from land to ocean through rivers. In: J.M. Martin, J.D. Burton and D. Eisma (eds), River Inputs to Ocean Systems. UNEP and UNESCO, Switzerland, pp. 33-51.

Benninger, L.K., 1978. ^{210}Pb balance in Long Island Sound. In: M.J. Drake and J.R. Holloway (eds), Geochim. et Cosmochim. Acta, 42 (6)pp. 1165-1174.

Bewers, J.M. and Yeats, P.A., 1981. Behaviour of trace metals during estuarine mixing. In: J.M. Martin, J.D. Burton and D. Eisma (eds), River Inputs to Ocean Systems. UNEP and UNESCO, Switzerland, pp. 103-115.

Bieri, R., Bricker, O., Byrne, R., Diaz, R., Helz, G., Hill, J., Huggett, R., Kerhin, R., Nichols, M., Reinharz, E., Schaffner, L., Wilding, D. and Strobel, C., 1982. Toxic substances, Part III. In: E.G. Macalaster, D.A. Barker and M. Kasper (eds), Chesapeake Bay Program Technical Studies: A Synthesis. U.S.E.P.A., Washington, DC, pp. 263-375.

Biggs,R.B., Sharp, J.H. and Howell, B.A., 1984. Suspended sediments. In: J.H. Sharp (ed.), The Delaware Estuary : Research as Background for Estuarine Management and Development. University of Delaware Sea Grant College Program, Newark, DE. DEL-SG-03-84, pp. 107-116.

Boyle, E., Collier, R., Dengler, A.T., Edmond, J.M., Ng, A.C. and Stallard, R.F., 1974. On the chemical mass balance of estuaries. Geochim. Cosmochim. Acta, 38: 1719-1728.

Bradford, W.L. and Luoma, S.N., 1980. Some perspectives on heavy metal concentrations in shellfish and sediment in San Francisco Bay, California. In: R.A. Baker (ed.), Contaminants and Sediments, Vol. 2. Ann Arbor Science Publishers, Inc., Ann Arbor, MI., pp. 501-532.

Capper, J., Power, G and Shivers, F.R., 1983. Chesapeake waters, pollution, public health and public opinion 1607-1972. Tidewater Publications, Centerville, MD, 201 pp.

Church, T.M., Tramontano, J.M. and Murray, S., 1984. Trace metal fluxes through the Delaware Estuary. Int'l. Council for Exploration of Sea (ICES) Symposium "Contaminant Fluxes through the Coastal Zone", Nantes, France, May, 1984.

Cifuentes, L.A., 1982. The character and behavior of organic nitrogen in the Delaware Estuary salinity gradient. M.S. Thesis, University of Delaware, Newark, DE, 81 pp.

Clarke, F.W., 1916. The Data of Geochemistry, 3rd Edition. U.S. Geol. Surv. Bull., 616. Government Printing Office, Washington, DC.

Coleman, P.J., 1969. The Transformation of Rhode Island 1790-1860. Brown University Press, Providence, RI, 314 pp.

Conomos, T.J. and Peterson, D.H., 1977. Suspended-particle transport and circulation in San Francisco Bay : An overview. In: L.E. Cronin (ed.), Estuarine Processes. Academic Press, New York, NY, Vol. 2, pp. 82-97.

Degens, E.T. (ed.), 1982. Transport of carbon and minerals in major world rivers, Part. I. Proceedings of a workshop arranged by the Scientific Committee on Problems of the Environment (SCOPE) and the United Nations Environment Programme (UNEP) at Hamburg University, March 8-12, 1982. SCOPE/UNEP Sonderband Heft 52. Palaontologischen Institut der Universitat Hamburg.

Dugdale, R.C., 1976. Nutrient cycles. In: D.H. Cushing and J.J. Walsh (eds),The Ecology of the Seas. W.B. Saunders Company, Philadelphia, PA., pp. 141-172.

Edmondson, W.T., 1969. Eutrophication in North America, pp. 124-149. In: Eutrophication Causes, Consequences, Correctives. NAS, Washington, DC.

Elderfield, H., Luedtke, N., McCaffrey, R.J. and Bender, M., 1981a. Benthic flux studies in Narragansett Bay. Am. J. Sci., 281: 768-787.

Elderfield, H., McCaffrey, R.J., Luedtke, N., Bender, M. and Truesdale, V.W., 1981b. Chemical diagenesis in Narragansett Bay Sediments. Am. J. Sci., 281: 1021-1055.

Evans, D.W., Cutshall, N.H., Cross, F.A. and Wolfe, D.A., 1977. Est. Coast. Mar. Sci., 5: 71.

Farrow, G.E., Syvitski, J.P.M. and Tunnicliffe, V., 1983. Suspended particulate loading on the Macrobenthos in a highly turbid Fjord : Knight Inlet, British Columbia, pp. 273-288. In: D.C. Gordon, Jr. and A.S. Hourston (eds), Proceedings of the Symposium on the Dynamics of Turbid Coastal Environments. Can. J. Fish. Aquat. Sci., 40 (suppl. 1): 1-365.

Furnas, M.J., 1982. The dynamics of summer phytoplankton populations in Narragansett Bay. Ph. D. Thesis, University of Rhode Island, Kingston, RI.

Furnas, M.J., Hitchcock, G.L., and Smayda, T.J., 1976. Nutrient-phytoplankton relationships in Narragansett Bay during the 1974 summer bloom. In: M. Wiley (ed.), Estuarine Processes, Vol. I., Uses, Stresses and Adaptations to the Estuary. Academic Press, NY, pp. 118-131.

Garber, J.H., 1984. ^{15}N tracer study of the short-term fate of particulate organic nitrogen at the surface of coastal marine sediments. Mar. Ecol. Prog. Ser., 16: 89-104.

Gearing, J.N. and Gearing, P.J., 1983. Suspended load and solubility affect sedimentation of petroleum hydrocarbons in controlled estuarine ecosystems. Can. J. Fish. Aquat. Sci., 40(2): 54-62.

Gearing, P.J., Gearing, J.N., Pruell, R.J., Wade, T.L. and Quinn, J.G., 1980. Partitioning of n°2 fuel oil in controlled estuarine ecosystems. Sediments and suspended particulate matter. Environ. Sci. Technol., 14(9): 1129-1136.

Giblin, A.E., 1982. Comparison of the processing of elements by ecosystems. II. Metals. II-i-II-24. Ecosystems Research Report N° 21. Cornell University, Ithaca, NY.

Goldberg, E.D., Gamble, E., Griffin, T.T. and Koide, M., 1977. Pollution history of Narragansett Bay as recorded in its sediments. Est. Coast. Mar. Sci., 5: 549-561.

Gosselink,J.G. and Baumann, R.H., 1980. Wetland inventories: wetland loss along the United States coast. Z. Geomorph. Summ. 34: 173-187.

Graham, W.F., Bender, M.L. and Klinkhammer, G.P., 1976. Manganese in Narragansett Bay. Limnol. Oceanogr., 21: 665-673.

Heerden, I.L. van, Wells, J.T. and Roberts, H.H., 1983. River-dominated suspended-sediment deposition in a New Mississippi Delta, pp. 60-71. In: D.C. Gordon, Jr. and A.S. Hourston (eds), Proceedings of the Symposium on the Dynamics of Turbid Coastal Environments. Can. J. Fish. Aquat. Sci., 40 (suppl. 1): 1-365.

Hoffman, E.J. and Quinn, J.G., 1984. Hydrocarbons and other pollutants in urban runoff and combined sewer overflows. A Report submitted to NOAA Oceans Assessments Division, Grant NA8ORAD00047.

Hunt, C.D., Unpublished data.

Hunt, C.D., 1983. Variability in the benthic Mn flux in coastal marine ecosystems resulting from temperature and primary production. Limnol. Oceanogr., 28(5): 913-923.

Hunt, C.D. and Smith, D.L., 1983. Remobilization of metals from polluted marine sediments. Can. J. Fish. Aquat. Sci., 40 (2): 132-142.

Hurtt, A.C., 1978. The distribution of hydrocarbons in Narragansett Bay sediment cores. M.S. Thesis, University of Rhode Island, Kingston, RI.

Hurtt, A.C. and Quinn, J.G., 1979. Distribution of hydrocarbons in Narragansett Bay sediment cores. Environ. Sci. Technol., 13: 829-836.

Jaworski, N.A., 1981. Sources of nutrients and the scale of eutrophication problems in estuaries. In: B.J. Neilson and L.E. Cronin (eds), Estuaries and Nutrients. Humana Press, Clifton, NJ., pp. 83-110.

Kaul, L.W. and Froelich, P.N., Jr., 1984. Modeling estuarine nutrient geochemistry in a simple system. Geochimica et Cosmochimica Acta, 48: 1417-1433.

Kelly, J.R., 1983. Benthic-pelagic coupling in Narragansett Bay. Ph. D. Thesis, University of Rhode Island, Kingston, RI.

Kelly, J.R. and Nixon, S.W., 1984. Experimental studies of the effect of organic deposition on the metabolism of a coastal marine bottom community. Mar. Ecol. Prog. Ser., 17 (2): 157-169.

Kremer, J.N. and Nixon, S.W., 1978. A coastal Marine Ecosystem, Simulation and Analysis. Ecological Studies 24, Springer-Verlag, NY, 217p.

Likens, G.E., Mackenzie, F.T., Richey, J.E., Sedell, J.R., Turekian, K.K. and Pye, V.I., 1981. Flux of organic carbon by rivers to the oceans. Report of a Workshop at Woods Hole, Massachusetts, September 21-25, 1980. U.S. Dept. of Energy, Office of Energy Research, CONF-8009140-UC-11, Washington, DC.

Livingstone, D.A., 1963. Chemical composition of rivers and lakes. U.S. Geol. Survey Prof. Paper 440G: 1-64.

Loder, T.C. and Glibert, P.M., 1980. Nutrient variability and fluxes in an estuarine system. In: V.S. Kennedy (ed.), Estuarine Perspectives. Academic Press, NY. pp. 111-122.

Martin, J.M., Burton, J.D. and Eisma, D. (eds), 1981. River inputs to ocean systems. Proceedings of a SCOR/ACMRR/ECOR/IAHS/UNESCO/CMG/IABO/IAPSO Review and Workshop held at FAO Headquarters, Rome, Italy, from 26 to 30 March 1979. UNEP and UNESCO, Switzerland.

McCaffrey, R.J., Myers, A.C., Davey, E., Morrison, G., Bender, M., Luedtke, N., Cullen, D., Froelich, P. and Klinkhammer, G., 1980. The relation between pore water chemistry and benthic fluxes of nutrients and manganese in Narragansett Bay, Rhode Island. Limnol. Oceanogr., 25(1): 31-44.

McLoughlin, W.G., 1979. Rhode Island. W.W. Norton & Company, NY. 240 p.

Meade, R.H., 1981. Man's influence on the discharge of fresh water, dissolved material and sediment by rivers to the Atlantic coastal zone of the United States. In: J.M. Martin, J.D. Burton and D. Eisma (eds), River Inputs to Ocean Systems. UNEP and UNESCO, Switzerland, pp. 13-17.

Melosi, M.V., 1981. Garbage in the Cities, Refuse, Reform, and the Environment, 1880-1908. Texas A and M University Press, College Station, TX. 268 p.

Meybeck, M., 1979. Rev. Geol. Dyn. Geogr. Phys., 21 (3): 215-246.

Mueller, J.A., Gerrish, T.A. and Casey, M.C., 1982. Contaminant inputs to the Hudson-Raritan Estuary. NOAA Technical Memorandum OMPA-21. Boulder, CO.

Nixon, S.W., 1980. Between coastal marshes and coastal waters. A review of twenty years of speculation and research on the role of salt marshes in estuarine productivity and water chemistry. In: P. Hamilton and K.B. Mac Donald (eds), Estuarine and Wetland Processes. Plenum Publishing Corporation, NY, pp. 437-525.

Nixon, S.W., 1981. Remineralization and nutrient cycling in coastal marine ecosystems, In: B.J. Neilson and L.E. Cronin (eds), Estuaries and Nutrients. Humana Press, NJ, pp. 111-138..

Nixon, S.W., 1983. Estuarine ecology. A comparative and experimental analysis using 14 estuaries and the MERL microcosms. Final report to the U.S. Environmental Protection Agency, Chesapeake Bay Program, Grant N° X-003259-01.

Nixon, S.W., Kelly, J.R., Furnas, B.N. and Oviatt, C.A., 1980. Phosphorus regeneration and the metabolism of coastal marine bottom communities. In: K.R. Tenore and C.B. Coull (eds), Marine Benthic Dynamics. University of South Carolina Press, Columbia, SC, pp. 219-242.

Nixon, S.W. and Kremer, J.N., 1977. Narragansett Bay. The development of a composite simulation model for a New England estuary. In: C. Hall and J. Day (eds), Ecosystem Modeling in Theory and Practice. Wiley Interscience, NY, pp. 622-673.

Nixon, S.W. and Lee, V., 1979. Spatial and temporal pollution gradients in Narragansett Bay. International Council for the Exploration of the Sea. Warsaw Poland. Limited distribution mimeo.

Nixon, S.W. and Lee, V., in press. Wetlands and water quality. A regional review of recent research in the United States on the role of fresh and saltwater wetlands as sources, sinks, and transformers of Nitrogen, Phosphorus and various heavy metals. Report to the Waterways Experiment Station, U.S. Army Corps of Engineers, Vicksburg, MS.

Nixon, S.W., Oviatt, C.A. and Hale, S.S., 1976. Nitrogen regeneration and the metabolism of coastal marine bottom communities. In: J.M. Anderson and A. Macfadyen (eds). The Role of Terrestrial and Aquatic Organisms in Decomposition processes. Blackwell Scientific Publications, London, pp. 269-283.

Nixon, S.W. and Pilson, M.E.Q., 1984. Estuarine total system metabolism and organic exchange calculated from nutrient ratios : An example from Narragansett Bay. In: V.S. Kennedy (ed.), The Estuary as a Filter. Academic Press. NY, pp. 261-290.

Nixon, S.W., Pilson, M.E.Q., Oviatt, C.A., Donaghay, P., Sullivan, B., Seitzinger, S., Rudnick, D. and Frithsen, J., 1984. Eutrophication of a Coastal Marine Ecosystem- An experimental study using the MERL microcosms. In: M.J.R. Fasham (ed.), Flows of Energy and Materials in Marine Ecosystems. Plenum Publishing Corporation, NY, pp. 105-135.

Olsen, S., Pilson, M.E.Q., Oviatt, C.A. and Gearing, J.N., 1982. Ecological consequences of low, sustained concentrations of petroleum hydrocarbons in temperate estuaries. MERL, Graduate School of Oceanography, University of Rhode Island, Narragansett, RI, 30 pp.

Pilson, M.E.Q., 1985. On the residence time of water in Narragansett Bay. Estuaries, 8: 2-14.

Postma, H. and Dijkema, K.S., 1982. In H. Postma (ed.), Hydrography of the Wadden Sea:

Movements and Properties of Water and Particulate Matter, Report n° 2, A.A. Balkema, Rotterdam.

Santschi, P.H., Adler, D.M. and Amdurer, M., 1983. The fate of particles and particle-reactive trace metals in coastal waters : Radioisotope studies in microcosmsIn: Wong, Boyle, Bruland, Burton and Goldberg (eds), Trace Metals in Sea Water. Plenum Publishing Corporation, pp. 331-349.

Santschi, P.H., Li, Y.H., Adler, D.M., Amdurer, M., Bell, J. and Nyffeler, U.P., 1983. The relative mobility of natural (Th, Pb and Po) and fall out (Pu, Am, Cs) radionucleides in the coastal marine environment: results from model ecosystems (MERL) and Narragansett Bay. Geochimica et Cosmochimia Acta, 47: 201-210.

Santschi, P.H., Nixon, S., Pilson, M. and Hunt, C., 1984. Inputs and accumulations of sediments, trace metals (Pb, Cu), and hydrocarbons in Narragansett Bay. Est. Coast. Shelf. Sci. 19: 427-449.

Schindler, D.W., 1974. Eutrophication and recovery in experimental lakes: implications for lake management. Science, 184: 897-899.

Seitzinger, S.P., Nixon, S.W. and Pilson, M.E.Q., 1984. Denitrification and nitrous oxide production in a coastal marine ecosystem. Limnol. Oceanogr., 29(1): 73-83.

Shaw, S.P. and Fredine, G.C., 1956. Wetlands of the United States: Their extent and their value to waterfowl and other wildlife. U.S. Dept. of the Interior, Fish and Wildlife Service, Circular 39, Reissued 1971.

Smayda, T.J., 1973. The growth of Skeletonema costatum during a winter-spring bloom in Narragansett Bay, Rhode Island. Norw. J. Bot., 20: 219-247.

Smullen, J.T., Taft, J.L. and MacKnis, J., 1982. Nutrient and sediment loads to the tidal Chesapeake Bay system. A synthesis. U.S.E.P.A. Chesapeake Bay Program, Annapolis, MD, pp. 147-162..

Sundby, B., Silverberg, N. and Chesselet, R., 1981. Estuarine mobilization and export of manganese: An example from the St. Lawrence Estuary. In: J.M. Martin, J.D. Burton and D. Eisma (eds), River Inputs to Ocean Systems. UNEP and UNESCO, Switzerland, pp. 231-238.

Thomson, J., Turekian, K.K. and McCaffrey, R.J., 1975. The accumulation of metals in and release from sediments of Long Island Sound. In: L.E. Cronin (ed.), Estuarine Research, vol. I., Academic Press, NY, pp. 28-44.

Turekian, K.K., 1977. The fate of metals in the oceans. Geochimica et Cosmochimia Acta, 41: 1139-1144.

Walsh, J.J. and Dugdale, R.C., 1971. A simulation model of the nitrogen flow in the Peruvian upwelling system. Inv. Pesq., 35: 309-330.

Wood, L.B., 1982. The Restoration of the Tidal Thames. Adam Hilger, Ltd., Bristol, 202 p.

THE RÓLE OF DETRITUS AT THE LAND-SEA BOUNDARY

K.H. MANN

Department of Fisheries and Oceans, Marine Ecology Laboratory, Bedford Institute of Oceanography, P.O.Box 1006, Dartmouth, Nova Scotia, B2Y 4A2 (Canada)

ABSTRACT

Mann, K.H., 1986. The role of detritus at the land-sea boundary. In: P. Lasserre and J.M. Martin (eds), Biogeochemical Processes at the Land-Sea Boundary. Elsevier, Amsterdam.

Detritus derived from vascular plants (seagrasses, marsh grasses and mangroves) contains much structural material that is indigestible to most animals. Such detritus must undergo a lengthy period of fragmentation and colonization by bacteria and fungi before it constitutes nutritious food for animals. During the process, a large proportion of its carbon is oxidized to CO_2. On the other hand the algae (seaweeds, benthic microalgae and phytoplankton) form detritus that is directly usable by animals, though its nutritional value may be further enhanced by microbial colonization. As a result of this difference, rather a small proportion of the carbon production of salt marshes, seagrass beds and mangroves enters coastal food webs. Isotope studies indicate that most of the carbon in coastal animals is derived from algal sources, with minor contributions from vascular marine plants. The dissolved organic matter produced by both phytoplankton and macrophytes is readily taken up by bacteria and these provide an important food source for planktonic and benthic animals. Bacteria associated with detritus food webs are an important mechanism for conserving nitrogen in coastal ecosystems.

INTRODUCTION

Detritus is organic matter in the process of decomposition. It was once alive, but now is dead and the tissues have begun to disintegrate both by autolysis and under the influence of the enzymes secreted by microbes. Historically, the tendency was to regard decomposition as synonymous with remineralization, and little interest was shown in detritus as a potential source of food for the animals in coastal food webs. It is now known that the ecological role of detritus is twofold. If dead organic matter is left in contact with microbes but isolated from higher organisms it will eventually decompose completely, releasing nutrient materials that are available for new cycles of plant production. But if, at some intermediate stage of decomposition the mixture of dead organic matter and microbes is ingested by an animal, it may provide the basis for a productive food web. At the present stage of marine ecology the key questions are : What proportion of detritus generated at the land-sea boundary is destined to be mineralized by bacteria and fungi and what proportion enters the food web as a result of ingestion by animals ? How important is detritus production compared with live photosynthetic production which enters the food web by herbivory ?

My attention was first drawn to the importance of detritus when I was working in the River Thames about 25 years ago. A limnological colleague, reviewing primary productivity of plants in lakes, rivers and wetlands pointed out that there was an enormous production of plant material by rooted plants, but nothing seemed to be eating them and they were therefore a "dead-end" as far as the freshweter food chains leading to fish were concerned. This proved to be quite untrue in our studies of the River Thames. We followed the energy budgets of several species of fish, month by month throughout the year. In the winter months the only way those fish balanced their energy budgets was by ingesting large quantities of detritus from the bed of the river. Living in that detritus were a few invertebrates, mainly worms and insect larvae, but our energy budgets showed that the fish were also taking considerable amounts of energy and nutrients from the detritus.

Similar observations can be made in the coastal zone. As we move towards the coast from offshore we notice that with the shallowing of the water it becomes possible for attached plants, seaweeds and seagrasses to receive enough light to grow on the bottom. Hard bottoms are colonized by algae, which have a holdfast but no rooting system. Soft, depositing substrates are colonized by rooted flowering plants, notably the seagrasses. In the intertidal zone where it is rocky the algae continue to the upper limit of the tidal excursion, except in hot, arid climates where they may be excluded by desiccation, or in northern latitudes where the grinding of ice on the shore prevents their colonization. In sedimented intertidal areas saltmarshes tend to dominate in temperate climates while mangroves become more important in the tropics.

All these plant communities produce large amounts of new tissue, but with a few exceptions nothing much eats them. They tend to grow for a season then die off and become detritus. When we study the process closely we find that much of what is released is particulate and readily visible e.g. fragments of leaves or of algal fronds, but there is also a considerable amount of dissolved organic matter. The dead particles soon become colonized by bacteria and fungi, and bacteria also take up much of the released dissolved organic matter. For the purpose of ecological balance sheets we have to consider the production of particulate organic matter (POM) with attached microorganisms and the production of dissolved organic matter (DOM) plus the bacteria associated with it, if these can be identified. For completeness we also have to consider any dead organic matter added to the pool by animals, for example faecal pellets, dead bodies, and shed exoskeletons. In discussing the role of detritus in this symposium we are therefore considering the role of all dead organic matter and its associated microorganisms. We should not overlook the fact that in coastal waters, as in the open ocean, phytoplankton also produces dead organic matter.

WORK PRIOR TO 1970

Petersen and his colleagues in Denmark made the claim as early as 1918 that it was the extensive seagrass beds around the coast of Denmark that provided the primary production to support the coastal fisheries through a detritus food web (Petersen, 1918). In the 1930's there was a great reduction in eelgrass biomass from what was called "the wasting disease of eelgrass" (although the cause of the mortality is still in dispute, Rasmussen, 1977). When there was no consequent collapse in the fisheries, the idea of a connection between eelgrass productivity and fish productivity was dropped. In the 1960's there was a revival of interest in the ecological role of coastal macrophytes, following a detailed study of the fate of the productivity of a salt marsh at Sapelo Island, Georgia, USA. Teal (1962) made the first energy budget for such a system and showed that while only about 7% of the fixed energy passed to herbivores, about 45% was exported from the marsh as POM. Odum and de la Cruz (1967) showed that in a series from living Spartina leaves, through dead brown leaves to fine particulate detritus, there was a 2-3 fold increase in nitrogen content. They also found that the weight specific rate of oxygen consumption of small particles (< 65 µm) was almost an order of magnitude greater than that of large particles (> 240 µm). They attributed this to an increase in the relative amounts of microorganisms on the detritus, consequent on an increase in the ratio of surface area to volume as the particles became more fragmented. The increase in microorganisms was thought to explain the increase in nitrogen content, and it was pointed out that fine detritus derived from Spartina was a nutritious source of food for many filter feeding animals. In an elegant experiment, Newell (1965) showed that when the faecal pellets of the snail Hydrobia were held for 3 days their nitrogen content increased by an order of magnitude. The snails could be induced to re-ingest them, and when the material came out again the nitrogen content was again low, and could again be observed to increase when held in seawater for 3 days (Fig. 1). The changes in nitrogen content were, of course, attributed to colonization by microbes and it was suggested that in passing through the gut the Hydrobia digested the microbes but was unable to digest the main contents of the faecal pellet.

Marine macrophytes, especially the flowering plants (seagrasses, marsh grasses and mangroves) have a high content of structural material such as cellulose, which most animals cannot digest. However, by the end of the 1960's mechanism had been found for transferring the production of macrophytes to invertebrates and hence to higher levels in the food chain. It was postulated (for review, see Mann, 1972 a) that refractory macrophyte tissue was colonized by fungi and bacteria, and that the invertebrates consuming the detritus digested the coating of bacteria and fungi. Stated simply, the theory was that invertebrates don't eat macrophyte detritus, they eat microbes which eat the detritus. This process is analogous to the situation in ruminant animals : they can't digest grass and leaves, the microbes in the rumen do it for them, and the ruminants digest the microbes and the soluble products of their enzyme activity. In the marine

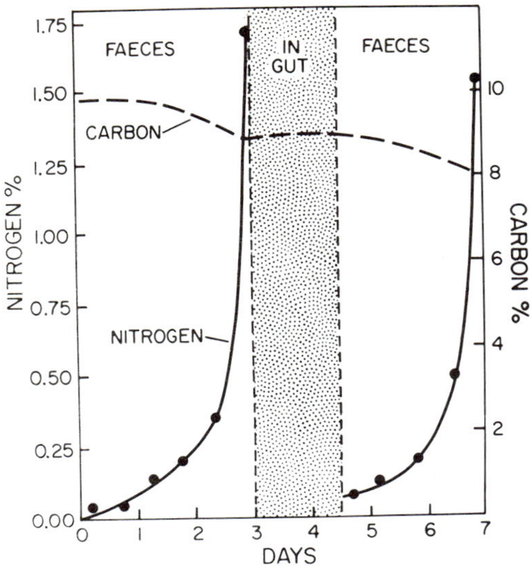

Fig. 1. Changes in the carbon and nitrogen content of faeces of the snail Hydrobia vulvae, (i) during 3 days in sea water, (ii) during passage through the gut and (iii) during a second period of exposure to sea water. From Newell (1965).

situation the microorganisms are usually in the environment rather than in the gut, but perform the same role.

With such a mechanism for trophic transfer available, people began to link the productivity of salt marshes, seagrass beds and kelp beds with the productivity of the coastal fisheries (e.g. Korringa, 1973, McHugh, 1976). Day et al. (1973) estimated that for Barataria Bay, Lousiana, 30% of the macrophyte production was exported to the Gulf of Mexico ; Valiela et al. (1978) found that 40% of the net annual production of the Great Sipiwisset marsh was exported as particulate matter to coastal waters ; Kikuchi (1971) documented the role of seagrass beds in Japan as feeding grounds for commercially important fish stocks ; Field et al. (1977) showed how kelp production off South Africa is used by mussels and other invertebrates which in turn support a commercially important stock of lobsters.

RATES OF DECOMPOSITION OF MATERIAL FROM DIFFERENT SOURCES

The processing of macrophyte detritus from the living conditions to the finest particles proceeds at different rates for material of different origins. Early attempts to investigate this took the form of exposing samples of material to the natural environment in replicate coarse-meshed nets or bags, and plotting the time course of disappearance. Absolute rates of disappearance depend on temperature, water movement and the ability of invertebrates to enter the bags, but under comparable conditions algae decompose most rapidly, followed by marsh grasses, while seagrasses appear most resistant to decay (Odum and de la Cruz, 1967 ; Wood et al., 1967 ; Zieman, 1968). Harrison and Chan (1980) showed that Zostera leaves that had been dead up to 2 weeks contained a water-soluble substance which inhibited bacteria and algae. They speculated that this substance might delay the decomposition process. Valiela et al. (1979) showed that newly dead Spartina contains considerable amounts of ferulic and coumaric acids, which are distasteful to amphipods and snails, and might therefore be expected to slow the process of shredding and grinding, which otherwise serves to accelerate the colonization by microbes. When we also consider that, compared with the algae, the vascular marine plants tend to have more structural material such as cellulose which is known to be resistant to the digestive enzymes of all but a few specilized organisms, we can begin to understand why decomposition is slower in vascular plants than in the algae.

The colonization of marine macrophytes by bacteria and fungi begins while they are still growing (Gesner et al., 1972 ; Meyers, 1974 ; Laycock, 1974) but the breaking up of the material into finer and finer particles by autolysis, water movement and particularly by the shredding and grinding of invertebrate, greatly increases the surface area and hence the populations of microbes (Fenchel, 1970). The microbes tend to find in the detritus an abundant source of carbon but limiting amounts of nitrogen and phosphorus. They are capable of taking these nutrients from the surrounding water, so that rates of decomposition are a function of the supply of N and P (Fenchel, 1977). Experiments have also shown that decomposition is faster in the presence of bacterial predators such as ciliates, rotifers, nematodes ans small crustaceans (Fenchel and Harrison, 1976). Their movements probably break down diffusion gradients of oxygen and nutrients around bacteria, hence stimulating bacterial metabolism and decomposition of the plant material. Larger grazers such as snails and amphipods probably have the same effect.

STUDIES ON THE NUTRITIONAL QUALITY OF DETRITUS

In spite of all the careful work on the colonization of detritus by microbes, and of the factors that influence the activity of those microbes, there has been a need for an integrated study of the whole process of detritus formation, ingestion by animals and production of new animal tissue. K.R. Tenore has explored in depth the conversion of dead plant tissue to animal tissue for one species of detritivore, the small polychaete worm Capitella capitata. Using standard cultures of the worm held under controlled conditions, he fed them with detritus of known origin and measured their growth. He found that two important properties of the detritus appeared to control the efficiency of its utilization : nitrogen content and "available caloric content", defined as that portion of the caloric content hydrolyzed by normal hydrochloric acid under standard conditions. Vascular plant detritus normally has a low nitrogen content (< 3%) and a low content of available calories (< 23% of total calories) because of its high content of structural materials unavailable to the animals. The algae commonly have both a higher nitrogen content and a higher content of available calories. At nitrogen contents below 3.5% the efficiency of incorporation of detritus into worm tissue was proportional to the nitrogen content, so that nitrogen appeared to be the limiting factor. At nitrogen levels about 3.5% the nitrogen was no longer limiting, and the efficiency of conversion to worm tissue was correlated with the available caloric content (Tenore, 1977, 1981). For example, naturally grown Spartina is normally rich in structural materials and proportionately low in available caloric content (13%). By growing Spartina under artificial conditions Tenore (1983) was able to boost the content of available calories to 44%. The efficiency of incorporation into Capitella tissue increased dramatically.

The proposed explanation of the difference between vascular plant detritus and seaweed detritus in efficiency of incorporation into Capitella was as follows. Vascular plant detritus, having a low content of nitrogen and available calories, must undergo a lengthy period of fragmentation and colonization by microbes before it is usable by Capitella. During that process, much of the tissue is mineralized by the microbes. Seaweed detritus needs a minimum of such conditioning. Much of it can be used directly by the animals, and the remainder is fairly rapidly coated with microorganisms and rendered palatable to the animals.

Two experiments tended to confirm this hypothesis. Tenore and Hanson (1980) prepared detritus from marsh grass (Spartina), seaweed (Gracilaria) and periphyton. Some was frozen soon after preparation, but other samples were aged in the presence of microbes for varying periods before being frozen. Capitella was then grown in each type of detritus. Measurements were made of the rate of oxidation of the detritus, the microbial biomass, and net incorporation into Capitella (Fig. 2).

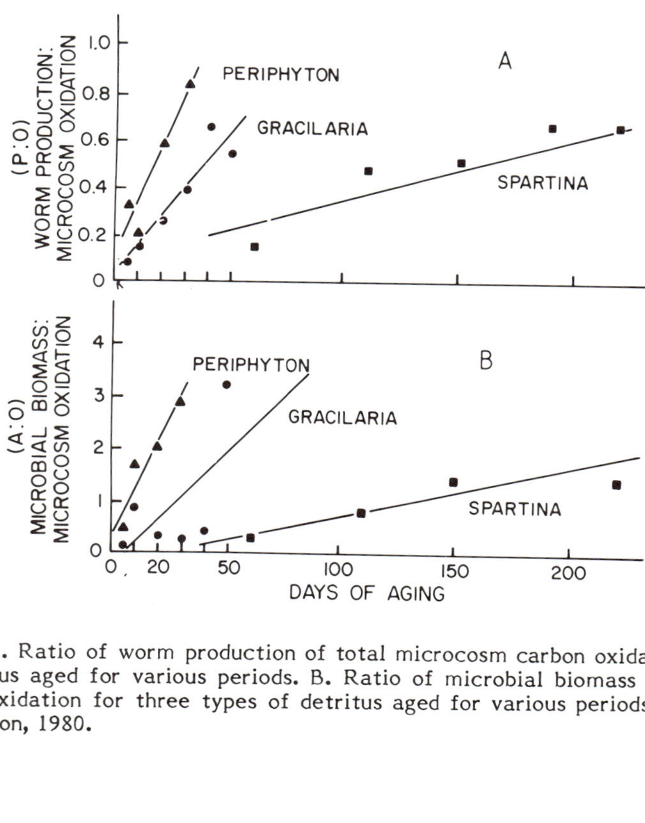

Fig. 2. A. Ratio of worm production of total microcosm carbon oxidation for three types of detritus aged for various periods. B. Ratio of microbial biomass to total microcosm carbon oxidation for three types of detritus aged for various periods. Both from Tenore and Hanson, 1980.

The Spartina oxidized more slowly than the algae, and in each case the efficiency of conversion of detritus to microbial biomass, and of conversion of detritus to Capitella biomass, (both on a daily basis) increased with the age of the detritus.

Findlay and Tenore (1982) succeeded in labelling plant material and microbes on detritus independently. They prepared three kinds of detritus : (i) with plant material labelled with ^{15}N, and microbes unlabelled ; (ii) with plant material unlabelled, but microbes labelled with ^{15}N ; and (iii) with both labelled. They did this for highly vascular material (marsh grass) and for seaweed. The worms feeding on vascular plant detritus were found to have obtained most of their ^{15}N from the microbes colonizing the surface, while worms feeding on the seaweed were found to have obtained most of their ^{15}N from the dead plant material.

Putting all these results together, it is clear that vascular plant detritus when newly formed is a poor source of nutrition for the worms. Only after a lengthy period of conditioning in the environment does it become a good source of nitrogen, carbon and calories. But during that conditioning period the microbes are oxidizing the plant tissues, and by the time it has become nutritious a good proportion has already been mineralized. Algal detritus, on the other hand, is relatively nutritious when newly formed, and is rapidly colonized by microbes so that it reaches peak condition in a few days. The net result is that algal material is used much more efficiently in detrital food webs than is marsh grass, seagrass or mangrove material. We must be careful not to generalize too broadly from work with one species of worm, but limited work with other species (e.g. Nephtys, Tenore et al., 1977 ; nematodes, Findlay, 1982) suggests that other parts of the food web have similar responses.

WORK WITH CARBON ISOTOPE RATIOS

a) Salt marshes

It has been found that vascular plants divide into two groups with respect to their stable carbon isotope ratios. Those with a C_3 pathway of photosynthesis have a $\delta^{13}C$ value of -24 to -34°/oo while C_4 plants have less negative values, -6 to -19°/oo, and algae have intermediate values of -12 to 23°/oo (Smith and Epstein, 1971). Animals feeding predominantly on one food source have been found to have a similar $\delta^{13}C$ value to that of their food. Haines (1976) and Haines and Montague (1979) looked at the carbon isotope values of a variety of plants and animals living in a Spartina marsh at Sapelo Island, Georgia (Fig. 3). They found that the Spartina and other grasses had $\delta^{13}C$ values of -12.3 to -13.6°/oo. Other vascular plants such as Juncus were -22.8 to -26.0°/oo while benthic diatoms were -16.2 to -17.9°/oo. The POM in the creeks had values in the range -19.8 to -22.8°/oo, suggesting that it was derived quite largely from phytoplankton (-20 to -22°/oo) rather than Spartina. Marsh snails and insects had $\delta^{13}C$ values close to those of Spartina, but deposit-feeding crabs reflected the ratios of the benthic diatoms. Mud snails and filter-feeding bivalves showed ratios similar to those of benthic and pelagic algae. In the subtidal regions of the creeks (Hughes and Sherr, 1983) mud crabs and two kinds of fish (mummichogs and mullet) had values indicating a sizeable component of Spartina detritus,

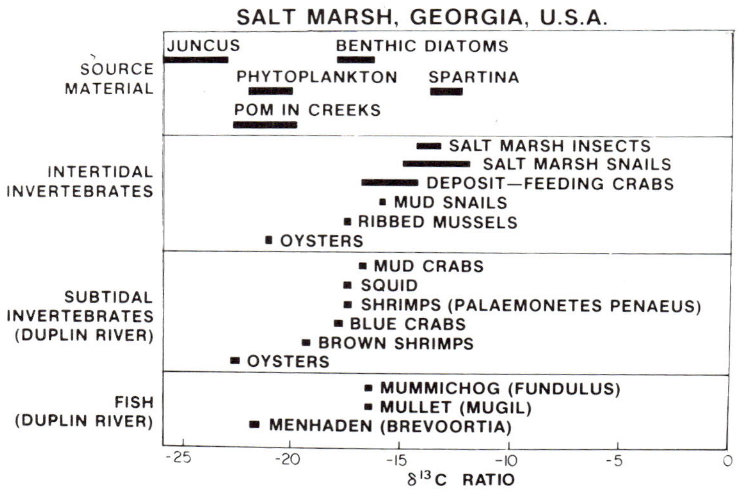

Fig. 3. Diagram of the stable carbon isotope ratios of organisms and POM in a Spartina marsh at Sapelo Island, Georgia. Data from Haines (1976) and Haines and Montague (1979).

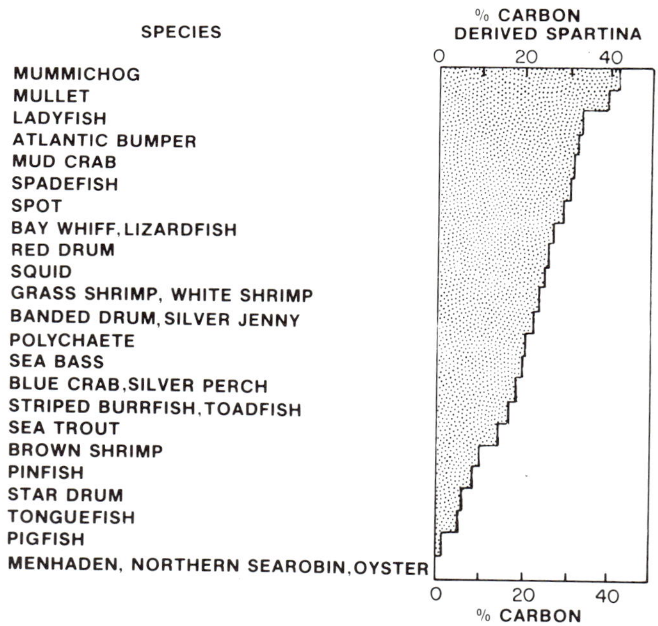

Fig. 4. Estimated percentages of carbon derived from Spartina for 30 invertebrate and fish species in a Georgia salt marsh estuary. Data from Hughes and Sherr (1983).

while brown shrimp, menhaden and oysters were apparently more dependent on the phytoplankton food web. These studies and others on salt marsh systems (reviewed in Fry and Sherr, in press) clearly indicate that even the fauna living in the creeks within the salt marshes derive only a minor part of their nourishment from salt marsh detritus (Fig. 4).

b) Seagrass beds

Analogous studies have been made in the vicinity of seagrass meadows. Thayer et al. (1978) found that eelgrass had ratios of -10.6 to -12.2°/oo, while the epiphytes were about -16°/oo. POM in suspension was much more negative than the eelgrass or its epiphytes, in the range -19 to 22°/oo and even the sediment in the eelgrass bed was about -19°/oo, indicating that phytoplankton (usually around -22°/oo) was making a major contribution to suspended and sedimented POM. The invertebrates and fish spanned the range -13.1 to -18.9°/oo, which could be explained by varying contributions from Zostera, epiphytes and plankton (Fig. 5). They concluded that "seagrass carbon plays a significant but not dominant role in the food web of this Zostera bed".

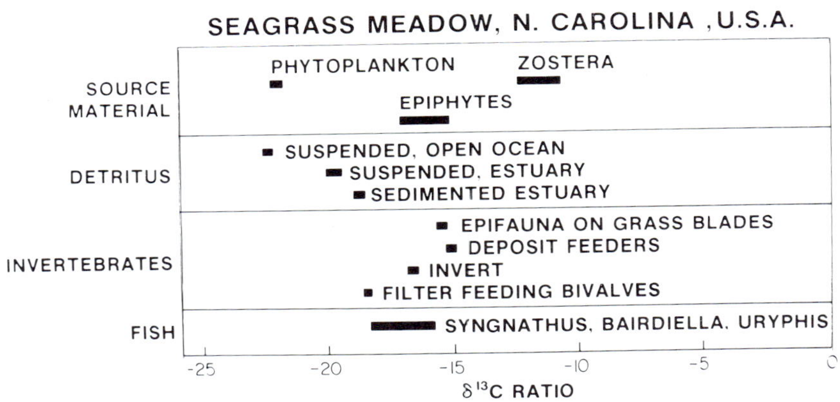

Fig. 5. Diagram of stable carbon isotope ratios in a seagrass meadow in North Carolina. Data from Thayer et al. (1978).

Similar results were obtained for Izembek Lagoon, in Alaska by McConnaughey and McRoy (1979). They concluded that benthic invertebrates such as the sea star Evasterias and the clam Macoma derive about 25% of their carbon from eelgrass, the shrimp Crangon dalli about 17%, the worm Nephtys caeca about 7%, while fish, depending whether they are resident or migratory derived from one third down to almost none of their carbon from the eelgrass. In arriving at these figures the authors took into account the possibility that changes in ^{13}C content occur in the food web, because ^{13}C is respired at a slower rate than ^{12}C. On the other hand lipid storage apparently enriches an animal in ^{12}C. The corrections required to allow for these shifts are still the subject of debate (reviewed in Fry and Sherr, in press) so that the percentages given above are quite approximate. Nevertheless, these and other studies make it clear that food webs in the vicinity of salt marshes and seagrass beds are by no means primarily dependent on detrital carbon.

This does not mean that we should now dismiss as unimportant the role of macrophyte detritus in estuarine and coastal food webs. Fry and Parker (1979) compared the stable carbon isotope ratios of animals from a Texas seagrass meadow with the ratios in comparable animals taken offshore. The polychaete worm Diopatra cuprea occurred in both situations and those taken from the seagrass meadows were up to 8.3°/oo enriched in ^{13}C relative to those from offshore. Fish and shrimp were enriched on average by 3.3°/oo and 5.1°/oo respectively. Clearly the seagrass detritus was making a significant contribution to the food web. Similarly, in the Caribbean, Fry et al. (1982) found a 4-6°/oo enrichment in ^{13}C in fish taken from seagrass meadows and coral reefs, relative to those taken offshore, and in the Torres Strait (Fig. 6) there were enrichments up to 8°/oo in comparable groups of animals from the seagrass beds compared with offshore, all indicating a substantial uptake of detrital carbon from the seagrasses (Fry et al., 1983).

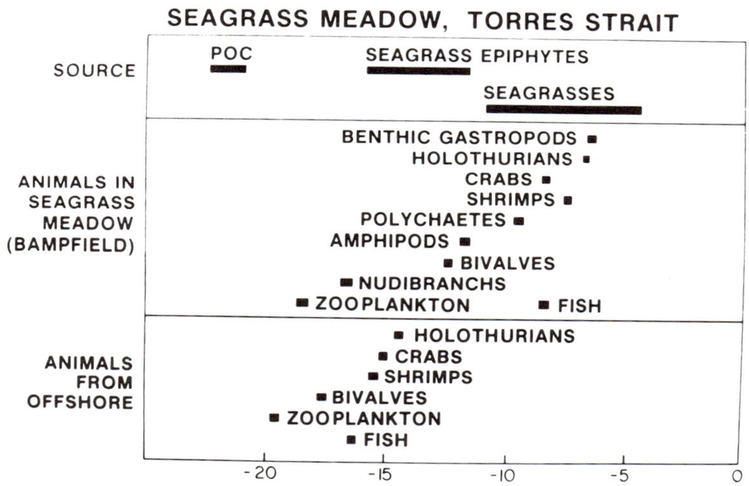

Fig. 6. Distribution of stable carbon isotope ratios in a seagrass meadow in the Torres Strait, compared with animals from offshore. Data from Fry et al., 1983.

When Fry (1984) used the carbon isotope technique to study the utilization of macrophyte carbon in a Florida bed of Syringodium he found that most of the fauna had ratios characteristic of the algae epiphytic on the seagrass blades. In this situation the productivity of the epiphytic algae was as high, or higher than that of the seagrass, and the seagrass detritus appeared to be carried away from the seagrass beds by strong tidal action.

c) Mangroves

Rodelli et al. (1984) recently reported the first stable carbon isotope study of a mangrove community. They obtained good separation of the primary producers, with benthic algae at -18.7 ± 2.2°/°°, phytoplankton at 21.0 ± 0.3°/°°, and mangroves at 27.1 ± 1.2°/°°. Twenty-nine of the 51 species of animals examined had ratios more negative than the algae, indicating uptake to mangrove carbon. These included some commercially important species such as bivalves and shrimps. Animals collected from offshore showed no evidence of uptake of mangrove carbon.

d) Algal seaweeds

On the whole, the stable carbon isotope ratio method has not proved to be a useful approach to tracing pathways of utilization of seaweed detritus. In part this is due to the close similarity between the isotope ratios of the seaweeds and those of the phytoplankton, but there is an additional problem that a particular blade of a single macrophyte may show a considerable variation from one part to another. Thus Stephenson et al. (1984) found that on a blade of the kelp Laminaria longicruris isotope ratios varied spatially from -12 to -20°/°°, and that on top of this was an irregular and apparently unpredictable seasonal variation. Clearly, this is no basis for careful discrimination between kelp detritus and phytoplankton (about -21°/°° with a range of -18 to -24°/°°, Fry and Sherr, in press). The question of the extent of utilization of kelp detritus in food webs has been approached using other techniques.

STUDIES OF FOOD WEBS BASED ON KELP DETRITUS

The seaweeds known as kelps and giant kelps (Laminariales) are built in such a way that they produce detritus more or less continuously. All are divisible into holdfast, stipe and blade (Fig. 7) and the blades normally have a growth zone at the base and a zone of erosion near the tip. They have been likened (Mann, 1972b) to "moving belts of tissue". Growth at the base compensates for erosion at the tip, and in the process large amounts of detritus, both particulate and dissolved, are released into the water. This detritus is readily colonized by bacteria (Laycock, 1974) and appears to be less resistant to digestion by animals than the detritus derived from vascular plants.

A group at Cape Town under the leadership of Dr. John Field has made an intensive study of the production and fate of tissue in a kelp bed dominated by Ecklonia maxima and Laminaria pallida. This system produces about 600 g C m^{-2} yr^{-1} as particulate organic matter and about 300 g C m^{-2} yr^{-1} as dissolved organic matter. At the same time, about 500 g C m^{-2} yr^{-1} is fixed by the phytoplankton (Newell and Field, 1983).

The main consumers are filter-feeders : mussels, sponges, and ascidians. Between them, they are capable of filtering a volume equal to all the water in the kelp bed once a day. Seiderer et al. (1982) showed that the mussels had enzyme systems capable of digesting the carbohydrates contained in the kelps, and Stuart (1982) showed that the mussels could grow well on a diet of kelp detritus. By an ingenious double-labelling technique Stuart et al. (1982) showed that the mussels in the kelp bed were actually incorporating carbon from the kelp debris into their tissues, and were not just removing the bacteria from its surface. Calculations showed that when the kelp bed system is "closed", with a minimal import and export, the combined production of kelp and phytoplankton fell a little short of the carbon requirements of the consumers.

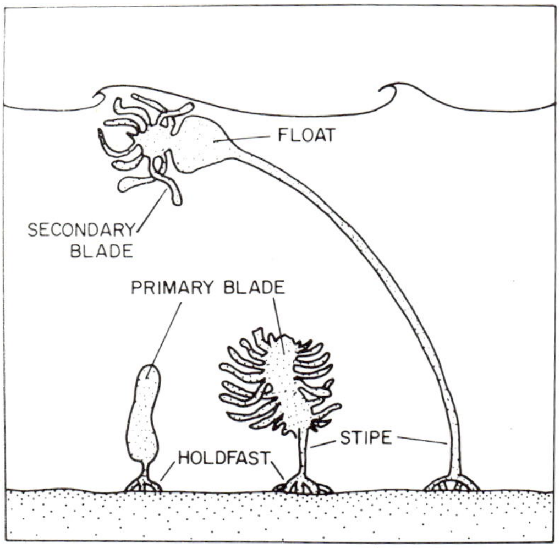

Fig. 7. Diagram of three main types of blade : i) with simple blade, e.g. Laminaria longicruris ; ii) with secondary blades, e.g. Ecklonia radiata ; iii) with elongated stipe, secondary blades and a float, e.g. Ecklonia maxima.

In practice, there were alternating periods of upwelling, when the consumers received mainly kelp detritus, and downwelling, when they received mainly phytoplankton imported from offshore. In a computer simulation Wulff and Field (1983) showed that kelp POM provides up to 70% of the food of filter feeders under upwelling conditions and 10-30% of the food under downwelling conditions. The relative importance of phytoplankton and detritus to the animals depends on the relative frequency of upwelling and downwelling conditions, and on the rates of water exchange in the kelp beds (Fig. 8). In the model, using a "realistic pulsing regime simulating conditions in the southern Benguela region" consumer biomass increased during winter and decreased during the summer upwelling season. Faecal production fed back into the system and was re-used by detritus feeders, thus providing a continuous supply of detritus even under downwelling conditions. The main predator in the system is the rock lobster Jasus lalandii, and this had the effect of dampening fluctuations in filter feeder biomass. We may note in passing that a high proportion of the world's lobster stocks are found in and around kelp beds where they rely on detritus-based food webs for a good proportion of their nutrition.

The Cape Town study also included consideration of the fate of kelp cast up on the beach by storms. This is not a trivial amount, it was estimated at 1200-1800 tonnes per km of shore line per year (Koop and Field, 1980). Griffiths and Stenton-Dozey (1981) showed that the main invertebrate colonists are amphipods, isopods and insects, their faeces being a major component of the pile few days after it is deposited on the shore. Using an ingenious series of field experiments, Koop et al. (1982a and b) found that the invertebrates consumed 74% of the kelp within a 8-day period while bacteria consumed 26%.

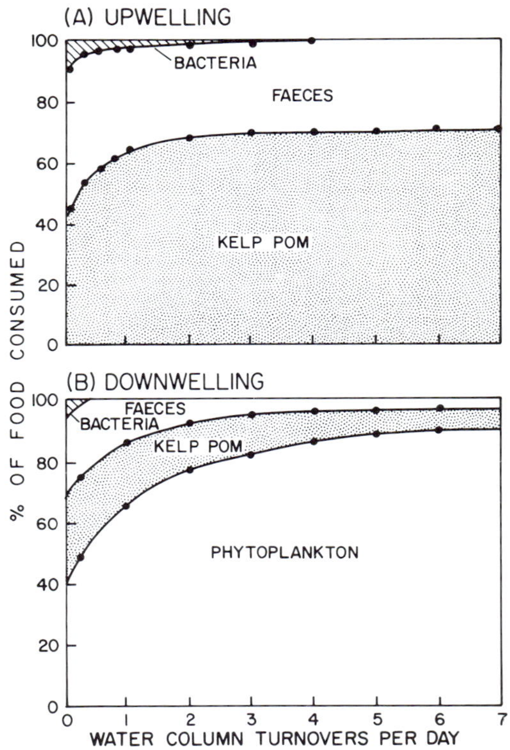

Fig. 8. Estimates of percentage of food consumed by kelp bed filter feeders that is derived from phytoplankton, kelp, faeces and bacteria. From the simulation model of Wulff and Field (1983).

Invertebrate faeces amounted to 67% of the initial kelp carbon, and these were in turn metabolized by bacteria in the sand. Overall, 100 g C of algal biomass yielded 23 - 28 g C of bacterial biomass.

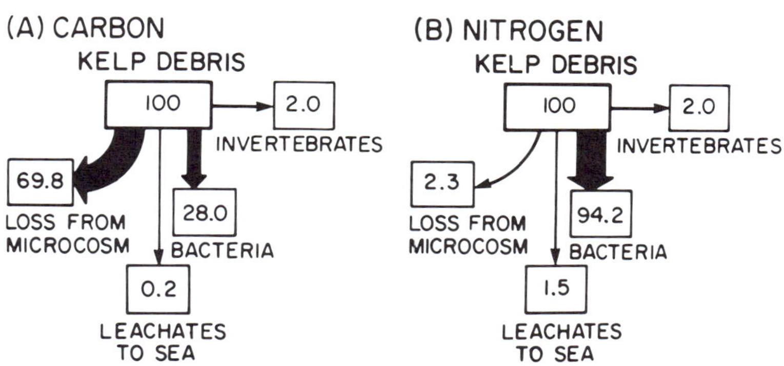

Fig. 9. Mass balances of (a) Carbon and (b) Nitrogen in a pile of decomposing kelp, Ecklonia maxima. Figures are g carbon or nitrogen derived from 100 g carbon or nitrogen kelp. Note that 69.8% of the carbon is lost from the microcosm but only 2.3% of the nitrogen.

When the nitrogen budget of this system was drawn up, it was found that the bacterial biomass incorporated 94% of the nitrogen present in the deposited kelp, even though it contained only 28% of the carbon (Fig. 9). This led to the idea that bacteria play an important role in conserving nitrogen in a detritus system. We now see that three lines of work point in the same direction. Laboratory feeding experiments have shown that detritus derived from marsh grass, seagrass or mangrove has a high content of structural material that is unavailable to detritivorous animals until it has been processed for a long time by bacteria and fungi. During that period much of the material is mineralized by mirobial action. Macro- and micro-algae on the other hand, have a much lower content of indigestible material and are rather rapidly conditioned by the microbiota. These facts help explain why, even for animals living in a salt marsh or seagrass bed, the greater part of the carbon in their bodies appears to be derived from planktonic or benthic algae. In the one kelp bed system that has been studied in sufficient detail, it appears that the invertebrate filter feeders receive considerable amounts of phytoplankton carbon.

DETRITUS DERIVED FROM PHYTOPLANKTON AND PERIPHYTON

While marine macrophytes are quite characteristic of the land-sea interface, the same cannot be said for phytoplankton. Yet the findings from work with stable carbon isotopes are that most organisms living even in seagrass beds of saltmarsh creeks derive more than half their carbon from algae, presumably phytoplankton and benthic microalgae. The whole question of the relationship between phytoplankton, their grazers, dissolved organic matter, bacteria, and their grazers, and the apparently changeable equilibrium between POM and DOM is poorly understood. In nearshore waters where the events are further confounded by additions of POM and DOM from macrophytes, the situation is even more confused. After very detailed studies in the Straits of Georgia, British Columbia, Parsons (1975) proposed the kind of relationship shown in Fig. 10.

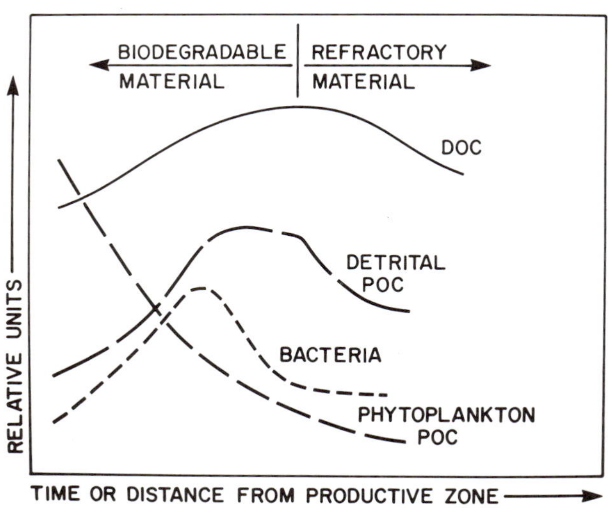

Fig. 10. Diagram showing peaks of DOC, POC and bacteria all occurring after a phytoplankton maximum. After Parsons (1975).

Phytoplankton cells give off dissolved organic matter in quantities that increase when nutrients are limiting or cells are senescent. Bacteria readily take up this material, but there are also several physical processes capable of converting DOM to POM. The bacteria themselves may give off considerable quantities of POM in the form of "rope-like aggregates" (Paerl, 1978). Detritus consisting of senescent phytoplankton and dead organic matter derived from a variety of sources, accumulates at the sediment-water interface in areas of low turbulence. To this is added faecal pellets from planktonic and benthic animals. Direct measurements with sediment traps and indirect estimates from measurements of community metabolism converge on the view that the input of detritus to the benthos ranges from under 100 g C m^{-2} yr^{-1} in less productive coastal areas to over 300 g C m^{-3} yr^{-1} in places where there is a major deposition of material from terrestrial plants or marine macrophytes. The details of the mechanisms by which detritus is processed in the benthos will be covered by other contributions to this symposium.

CONCLUSION

A key question identified in the introduction was : What proportion of detritus generated at the land-sea boundary is destined to be mineralized by bacteria and fungi, and what proportion enters the food web as a result of ingestion by animals ? The answer will, of course, be somewhat different in each locality, but some trends are beginning to be seen. Because detritus derived from vascular marine plants is more indigestible than detritus derived from seaweeds or microalgae, it is inherently probable that a larger proportion of it will be mineralized by microbes without being utilized in animal food webs. But almost all studies to date have indicated that even in areas of abundant macrophytes, algal carbon is an important component of all animal tissue. It looks as if detritus is a secondary source of nutrition for most coastal communities. It may well be that for animals that have access to living phytoplankton or benthic algae, these provide the main life support during the "bloom" seasons. Detritus is less seasonal in occurrence - vascular plant detritus resists decay for many months, and seaweed detritus is produced more or less continuously. It seems likely that detritus provides a reserve of food for times when fresh food is less plentiful. This was the tentative conclusion of Heinle and Flemer (1975) for zooplankton in the Patuxent River Estuary and Stuart, Head and Mann (in press) for amphipods in the Bay of Fundy.

REFERENCES

Day, J.A., Smith, W.G., Wagner, P.R. and Stowe, W.C., 1973. Community Structure and Carbon Budget of a Saltmarsh and Shallow Bay Estuarine System in Louisiana. Center for Wetland Resources, Louisiana State University.

Fenchel, T., 1970. Studies on decomposition of organic detritus derived from the turtle grass Thalassia testudinum. Limnol. Oceanogr., 15: 14-20.

Fenchel, T., 1977. Aspects of decomposition of seagrasses. In: C.P. Mc Roy and C. Hellferich (eds) Seagrass Ecosystems: A Scientific Perspective. Marcel Dekker, New York, pp.123-145.

Fenchel, T. and Harrison, P., 1976. The significance of bacterial grazing and mineral cycling for the decomposition of particulate detritus. In: J.M. Anderson and A. Macfadyen (eds), The Role of Terrestrial and Aquatic Organisms in Decomposition Processes. Blackwell Scientific Publications, Oxford, pp. 285-299.

Field, J.G., Jarman, N.G., Dieckman, G.S., Griffiths, C.L., Velimirov, B. and Zoutendyck, P., 1977. Sun, waves, seaweed and lobsters: The dynamics of a west-coast kelp bed. South African J. Sci., 73: 7-10.

Findlay, S.E.G., 1982. Effect of detrital nutritional quality on population dynamics of a marine nematode (Diploaimella chitwoodi). Mar. Biol., 68: 226-227.

Findlay, S.E.G. and Tenore, K.R., 1982. Nitrogen source for a detritivore: Detritus substrate versus associated microbes. Science, 218: 371-373.

Fry, G., 1984. $^{13}C/^{12}C$ ratios and the trophic importance of algae in Florida Syringodium filiforme seagrass meadows. Mar. Biol., 79: 11-19.

Fry, B., Lutes, R., Northam, M. and Parker, P.L., 1982. A $^{13}C/^{12}C$ comparison of food webs in Caribbean seagrass meadows and coral reefs. Aquat. Bot., 14: 389-398.

Fry, B. and Parker, P.L., 1979. Animal diet in Texas seagrass meadows : $^{13}C/^{12}C$ evidence for the importance of benthic plants. Estuar. Coastal Mar. Sci., 8: 499-509.

Fry, B., Scanlan, R.S. and Parker, P.L., 1983. $^{13}C/^{12}C$ ratios in marine food webs in the Torres Strait, Queensland. Aust. J. Mar. Freshwat. Res., 34: 707-715.

Fry, B. and Sherr, E.B., in press. $^{13}C/^{12}C$ measurements as indicators of carbon flow in marine and freshwater ecosystems. Contrib. Mar. Sci. .

Gessner, R.V., Goos, R.D. and Sieburth J. Mc N., 1972. The fungal microcosm of the internodes of Spartina alterniflora. Mar. Biol., 16: 269-273.

Griffiths, C.L. and Stenton-Dozey, J., 1981. The fauna and rate of degradation of stranded kelp. Estuar. Coastal and Shelf Sci., 12: 645-653.

Haines, E.B., 1976. Stable carbon isotope ratios in the biota, soils and tidal waters of a Georgia salt marsh. Estuar. Coast. Mar. Sci., 4: 609-616.

Haines, E.B. and Montague, C.L., 1979. Food sources of estuarine invertebrates analyzed using $^{13}C/^{12}C$ ratios. Ecology, 60: 48-56.

Harrison, P.G. and Chan, A.T., 1980. Inhibition of the growth of microalgae and bacteria by extracts of eelgrass, (Zostera marina) leaves. Mar. Biol., 61: 21-26.

Heinle, D.R. and Flemer, D.A., 1975. Carbon requirements of a population of the estuarine copepod Eurytemora affinis. Mar. Biol., 31: 235-247.

Hughes, E.H. and Sherr, E.B., 1983. Subtidal food webs in a Georgia estuary : $^{13}C/^{12}C$ analysis. J. Exp. Mar. Biol. Ecol., 67: 227-242.

Kikuchi, T., 1974. Japanese contributions on consumer ecology in eelgrass (Zostera marina L.) beds, with special reference to trophic relations and resources in inshore fisheries. Aquaculture, 4: 145-160.

Koop, K. and Field, J.G., 1980. The influence of food availability on the population dynamics of a supralittoral isopod Ligia dilatata Brandt. J. Exp. Mar. Biol. Ecol., 48: 61-72.

Koop, K., Newell, R.C. and Lucas, M.I., 1982a. Biodegradation and carbon flow based on kelp (Ecklonia maxima) debris in a sandy beach microcosm. Mar. Ecol. Prog. Ser., 7: 315-326.

Koop, K., Newell, R.C. and Lucas, M.I., 1982b. Microbial regeneration of nutrients from the decomposition of macrophyte debris on the shore. Mar. Ecol. Prog. Ser., 9: 91-96.

Korringa, P., 1973. The edge of the North Sea as a nursery ground and shellfish area. In: E.D. Goldberg (ed.), North Sea Science. M.I.T. Press, Cambridge, Mass, pp. 361-381.

Laycock, R.A., 1974. The detrital food chain based on seaweeds. I. Bacteria associated with the surface of Laminaria fronds. Mar. Biol., 24: 223-231.

Mann, K.H., 1972a. Macrophyte production and detritus food chains in coastal waters. Mem. Ist. Ital. Idrobiol., 29 suppl.: 353-383.

Mann, K.H., 1972b. Ecological energetics of the seaweed zone in a marine bay on the Atlantic coast of Canada. II. Productivity of the seaweeds. Mar. Biol., 14: 199-209.

McConnaughey, T. and McRoy, C.P., 1979. ^{13}C label identifies eelgrass (Zostera marina) carbon in an Alaskan estuarine food web. Mar. Biol., 53: 263-269.

McHugh, J.L., 1976. Estuarine fisheries : are they doomed ? In: M. Wiley (ed.) Estuarine Processes. Academic Press, New York, Vol. 1, pp. 15-27.

Meyers, S.P., 1974. Contribution of fungi to biodegradation of Spartina and other brackish marshland vegetation. Veröff. Int. Meeresforsch Bremerhaven, Suppl. 5: 357-375.

Newell, R.C., 1965. The role of detritus in the nutrition of two marine deposit feeders, the prosobranch Hydrobia ulvae and the bivalve Macoma balthica. Proc. Zool. Soc. Lond., 144: 25-45.

Newell, R.C. and Field, J.G., 1983. The contribution of bacteria and detritus to carbon and nitrogen flow in a benthic community. Mar. Biol. Letters, 4: 23-36.

Odum, E.P. and de la Cruz, A.A., 1967. Particulate organic detritus in a Georgia saltmarsh-estuarine ecosystem. In: G.H. Lauff (ed.), Estuaries. Amer. Assoc. for the Adv. of Sci. Washington, pp. 383-388.

Paerl, H.W., 1978. Microbial organic carbon recovery in aquatic ecosystems. Limnol. Oceanogr., 23: 927-935.

Parsons, T.R., 1975. Particulate organic carbon in the sea. In: J.P. Riley and G. Skirrow (eds) Chemical Oceanography. Academic Press, New York, Vol. 2, 2nd edn. pp. 365-383.

Petersen, C.J.G., 1918. The sea bottom and its production of fish food. A survey done in connection with the valuation of the Danish waters from 1883-1917. Rep. Danish. Biol. Sta., 25: 1-82.

Rasmussen, E., 1977. The wasting disease of eelgrass (Zostera marina) and its effects on environmental factors and fauna. In: C.P. McRoy and C. Hellfferich (eds), Seagrass Ecosystems : A Scientific Perspective. Marcel Dekker, New York, pp. 1-51.

Rodelli, M.R., Gearing, J.N., Gearing, P.J., Marshall, N. and Sasekumar, A., 1984. Stable isotope ratio as a tracer of mangrove carbon in Malaysian ecosystems. Oecologia, 61: 326-333.

Seiderer, L.J., Newell, R.C. and Cook, P.A., 1982. Quantitative significance of style enzymes from two marine mussels (Choromytilus meridionalis Krauss and Perna perna L.) in relation to diet. Mar. Biol. Letters, 3: 257-271.

Smith, R.N. and Epstein, S., 1971. Two categories of $^{13}C/^{12}C$ ratios for higher plants. Plant. Physiol., 47: 380-384.

Stephenson, R.L., Tan, F.C. and Mann, K.H., 1984. Stable carbon isotope variability in marine macrophytes and its implications for food web studies. Mar. Biol., 81: 223-230.

Stuart, V., 1982. Absorbed ration, respiratory costs and resultant scope for growth in the mussel Aulacomya ater (Molina) fed on a diet of kelp detritus of different ages. Mar. Biol. Letters, 3: 289-306.

Stuart, V., Field, J.G. and Newell, R.C., 1982. Evidence for absorption of kelp detritus by the ribbed mussel Aulacomya after using a new ^{15}C-labelled microsphere technique. Mar. Ecol. Prog. Ser., 9: 263-271.

Stuart, V., Head, E.J.H. and Mann, K.H., in press. Relative contributions of benthic diatoms and Spartina alterniflora (Loisel) detritus to the nutrition of the amphipod Corophium volutator (Pallas) on a Bay of Fundy mud flat. J. Exp. Mar. Biol. Ecol.

Teal, J.M., 1962. Energy flow in the salt marsh ecosystem of Georgia. Ecology, 43: 614-624.

Tenore, K.R., 1977. Growth of Capitella capitata cultured on various levels of detritus derived from different sources. Limnol. Oceanogr., 22: 936-941.

Tenore, K.R., 1981. Organic nitrogen and caloric content of detritus. I. Utilization by the deposit-feeding polychaete Capitella capitata. Estuar. Coastal and Shelf Sci., 17: 733-742.

Tenore, K.R., 1983. What controls the availability of detritus derived from vascular plants: organic nitrogen enrichment or caloric availability ? Mar. Ecol. Prog. Ser. 10: 307-309.

Tenore, K.R. and Hanson, R.B., 1980. Availability of detritus of different types and ages to a polychaete macroconsumer Capitella capitata. Limnol. Oceanogr., 25: 553-558.

Tenore, K.R., Tietjen, J.H. and Lee, J.J., 1977. Effect of meiofauna on incorporation of aged eelgrass, Zostera marina detritus by the polychaete Nephtys incisa. J. Fish. Res. Board Can., 35: 563-567.

Thayer, G.W., Parker, P.L., La Croix, M.W. and Fry, B., 1978. The stable carbon isotope ratio of some components of an eelgrass (Zostera marina) bed. Oecologia, 35: 1-12.

Valiela, I., Koumjian, L. and Swain, T., 1979. Cinnamic acid inhibition of detritus feeding. Nature, 250: 55-57.

Valiela, I., Koumjian, L., Swain, T., Teal, J.M. and Hobbie, J.E., 1979. Cinnamic acid inhibition of detritus feeding. Nature, 180: 55-57.

Wood, E.J.F., Odum, W.E. and Zieman, J.C., 1967. Influence of seagrasses on the productivity of coastal lagoons. In: Lagunas Costeras, un Symposia. Mem. Simp. Intern. Lagunas Costeras. UNAM-UNESCO, Mexico, D.F., pp. 495-502.

Wulff, F.V. and Field, J.G., 1983. Importance of different trophic pathways in a nearshore benthic community under upwelling and downwelling conditions. Mar. Ecol. Prog. Ser., 12: 217-228.

Zieman, J.C., 1968. A study of the growth and decomposition of seagrass Thalassia testudinum. Ms. Thesis, University of Miami.

SPATIAL AND TEMPORAL DISTRIBUTION OF BACTERIAL POPULATIONS IN MARINE SHALLOW WATER SURFACE SEDIMENTS

L.-A. MEYER-REIL

Institut für Meereskunde an der Universität Kiel, Marine Mikrobiologie, Düsternbrooker Weg 20, 2300 Kiel 1 (Federal Republic of Germany)

ABSTRACT.

Meyer-Reil, L.A., 1986. Spatial and temporal distribution of bacterial populations in marine shallow water surface sediments. In: P. Lasserre and J.M. Martin (eds), Biogeochemical Processes at the Land-Sea Boundary. Elsevier, Amsterdam.

In the paper presented, spatial and temporal aspects of the distribution of bacterial populations in marine shallow water surface sediments are discussed. Bacterial number and biomass as determined by epifluorescence microscopy turned out to be closely related to sediment properties such as grain size and organic matter content. For an "average" sediment, one may expect 10^9 cells per g of dry weight sediment with a corresponding biomass of 10 ug bacterial C per cell. Benthic bacterial populations underwent strong seasonal variations which could be traced back to the accumulation of organic material in the sediment following sedimentation events. In shallow water surface sediments bacterial communities were governed by diurnal rhythms obviously closely related to microphytobenthos primary production. Estimates of bacterial production using quite different approaches demonstrated that between 10 and 800 mg C m^{-2} d^{-1} were fixed by bacterial secondary production.

INTRODUCTION

Coastal sediments play an important role in nutrient regeneration for marine ecosystems. These processes are governed by the activity of benthic bacteria. Microbiological work in sediments has been concentrated on the investigation of nutrient cycles as well as on the spatial and temporal distribution of bacteria. In the studies of nutrient cycles, the activity of benthic bacteria was indirectly concluded from changes in the concentrations or turnover rates of inorganic and organic chemical parameters. Generally, corresponding information on the development of the bacterial populations themselves could not be included.

This paper summarizes information on the spatial and temporal distribution of benthic bacterial populations. Many examples discussed originate from microbiological investigations of sediments from the brackish water Kiel Bight (Baltic Sea, Federal Republic of Germany). Existing data and concepts illustrating the spatial and temporal distribution of bacteria in shallow water surface sediments are organized under the following topics :

1) Sediments as habitats for bacteria
2) Bacterial number and biomass
3) Seasonal development of bacterial communities
4) Diurnal fluctuations of bacterial populations
5) Estimates of bacterial production.

SEDIMENTS AS HABITATS FOR BACTERIA

Sediments represent a complex environment consisting of particles which are more or less densely packed and surrounded by interstitial water. In beach sediments which are extremely exposed to wave action and tidal activity more than 95 % of the bacteria are attached to particle surfaces (Meyer-Reil et al., 1978). Deeper sediments carry a much higher percentage of bacteria "free-floating" in the interstitial water (up to 50 % of the total number ; Weise and Rheinheimer, 1978 ; Weise and Rheinheimer, 1979).

Bacteria colonize only a small proportion of the available particle surface area (0.01 to 5 % ; range of the data obviously dependent upon the method used for the calculation of surface area ; cf. Hargrave, 1972 ; Rublee and Dornseif, 1978 ; Weise and Rheinheimer, 1978 ; DeFlaun and Mayer, 1983). Deep pores of wheathered feldspar and clay grains do not appear to be inhabited by bacteria (DeFlaun and Mayer, 1983). Bacteria preferentially

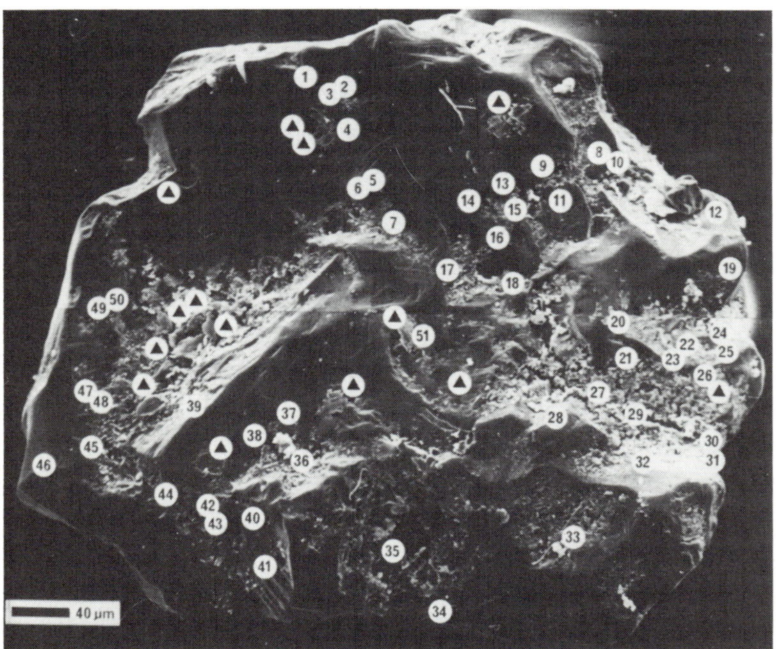

Fig. 1. Scanning electron micrograph of a 0.5 mm quartz grain. Note that organic material is concentrated in protected areas. The numbers on the quartz grain mark sites of specific investigation (for details cf. Weise and Rheinheimer, 1978).

colonize areas of low relief such as depressions and crevices of particles where the cells are protected against grazing and mechanical demages (Nickels et al., 1981 ; cf. Fig. 1). The habitation in these areas is an expression of preferential survival rather than preferential colonization as it could be demonstrated by DeFlaun and Mayer (1983).

The benthic bacterial flora is of a high diversity : rods, cocci and curved cells of different sizes. Single cells or small colonies are prevailing. Most of the bacteria are found in extracellular slime layers consisting of bacterial fibrous webs and mucus produced by diatoms (Weise and Rheinheimer, 1978 ; Moriarty and Hayward, 1982). Through the formation of this organic material of a considerable structural complexity, bacteria may influence the texture of sediments like it was demonstrated for fungi mycelium in sand dune soils (Clough and Sutton, 1978).

BACTERIAL NUMBER AND BIOMASS

As basic members of the food chain, benthic bacteria represent an important nutrient source for meio- and macrofauna. The distribution of bacterial number and biomass was shown to be closely related to sediment properties, from which organic material and grain size are probably most important.

Number and biomass as determined by epifluorescence microscopy

Beside scanning electron microscopy, epifluorescence microscopy permits a reliable estimate of number and biomass of bacteria in sediments. However, prior to counting bacteria have to be liberated from the particle surfaces. Different extraction techniques have been applied : treatment of sediments with surface active agents, homogenization (Meyer-Reil et al., 1978), and sonication (Weise and Rheinheimer, 1978 ; Ellery and Schleyer, 1984). From these techniques, sonication gave the most reliable results with the highest percentage of bacteria (approximately 95 % of the total number) liberated from the particle surfaces (Meyer-Reil, 1983). However, most recently this technique has been discussed with special regard to a possible destruction of bacterial cells (Ellery and Schleyer, 1984).

After sonication sediment samples are diluted, filtered onto prestained Nucleopore polycarbonate filters (pore size 0.2 µm) and stained with fluorescence dyes (e.g. acridine orange). Bodies with clear outline, bacterial shape and distinct fluorescence (orange or green) are counted as bacterial cells. For biomass determinations slides of characteristic microscopic fields can be prepared. The slides may be analysed by means of an image analyser which reduces the uncertainty to group the bacteria into arbitrary size classes by eye (Krambeck et al.,1981 ; Meyer-Reil, 1983). Using conversion factors, data can be extrapolated to biomass in terms of carbon (cf. Ferguson and Rublee, 1976 ; Cammen, 1982 ; Bakken and Olson, 1983).

Uncertainties arising from inadequacies of the sonication technique and the conversion factors are obviously minor as compared to the subjectivity involved in the counting procedure. However, up to now, beside scanning electron microscopy, epifluorescence microscopy is the only direct approach to gain information on bacterial number and biomass (cf. Fig. 2).

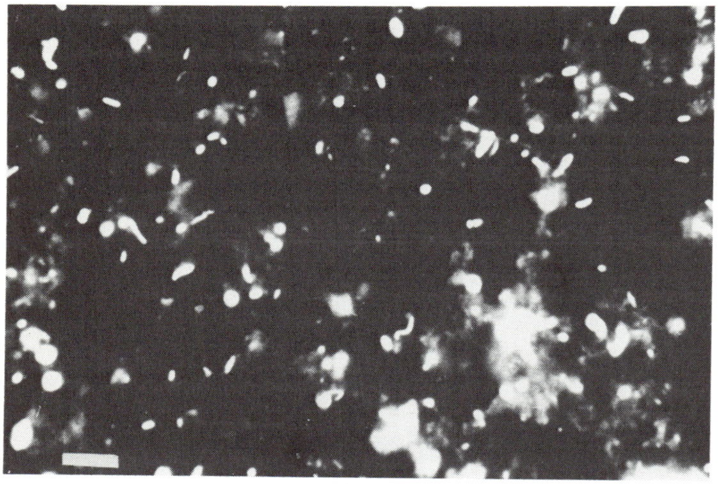

Fig. 2. Epifluorescence photograph of bacteria from a sandy sediment of the Kiel Bight (Baltic Sea ; FRG). The bacteria were liberated from the sediments by sonication. The sample was diluted and stained with acridine orange. Bar represents 3 um (from Meyer-Reil, 1984a).

Local variations of number and biomass

Data on local variations in bacterial number and biomass summarized from the literature are presented in Fig. 3 and 4. The data set comprises sediments from arctic, antarctic, boreal and tropical regions. Generally, bacterial numbers are in the range of 10^8 to 10^{11} cells per gram of dry weight sediment with a concentration of the data around 10^9 cells (cf. Fig. 3). Numbers turned out to be much more related to the type of sediment than to the region where the samples were taken. The lowest cell numbers were found in sandy sediments, the highest numbers in muddy sediments (cf. below). With increasing sediment depths, a slight decrease in bacterial numbers becomes obvious. However, even in the 11 m horizon of a sediment profile from the Antarctica, 10^9 cells per gram of sediment could be detected, only insignificantly less than in the surface horizon (Meyer-Reil, 1984 a).

Because of the time consuming procedure of size fractionation, much less data are available on bacterial biomass (Fig. 4). Generally, bacterial carbon is in the range of 1 to 10^3 µg per gram of dry weight sediment thus contributing significantly to the total benthic biomass as determined by ATP measurements (for data cf. Graf et al., 1982 ; Graf et al., 1983). Again, the lowest bacterial biomass values were found in sandy sediments, the highest values in muddy sediments.

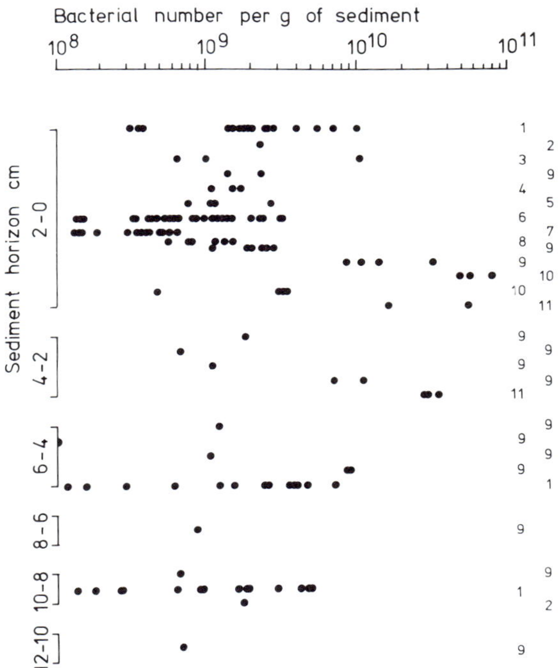

Fig. 3. Summary of data concerning bacterial numbers in sediments of different areas. Each of the dots represent one data point taken from the literature listed on the right panel : 1-Dale, 1974 ; 2-Kepkay et al., 1979 ; 3-Griffiths et al., 1978 ; 4-Meyer-Reil et al., 1981 ; 5-Meyer-Reil et al., 1978 ; 6-Meyer-Reil et al., 1980 ; 7-Meyer-Reil, 1981 ; 8-Weise and Rheinheimer, 1978, 1979 ; 9-Meyer-Reil unpublished data 10-Rheinheimer, unpublished data ; 11-Jones, 1980 ; 12-Moriarty, 1980.

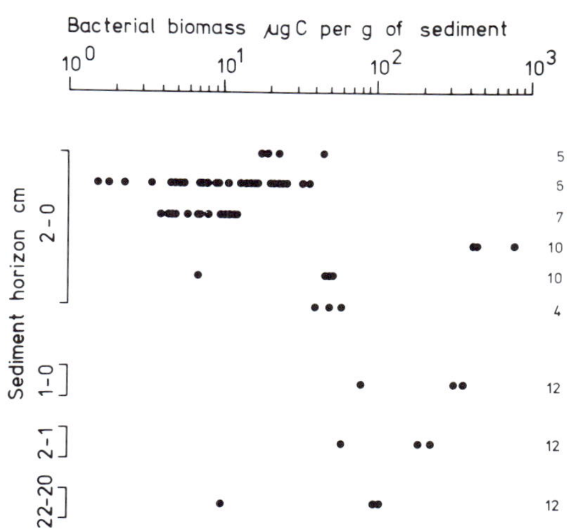

Fig. 4. Summary of data concerning bacterial biomass in sediments of different areas. For explanations see Fig. 3.

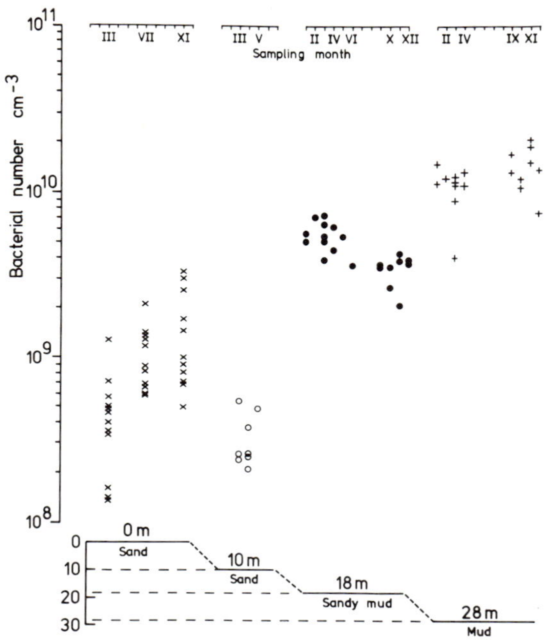

Fig. 5. Bacterial numbers per cm^3 of sediment on a hypothetical horizontal profile from sandy sediments (water depth 0, 10 m) to sandy-mud sediments (water depth 18 m) and to muddy sediments (water depth 20 m) of the Kiel Bight (Baltic Sea, FRG).

Relationship between number and grain size

A horizontal profil from beaches (water depth 0 m) into deeper waters (18, 28 m) of the Kiel Bight revealed an inverse relationship between benthic bacterial number and grain size of the sediment (Fig. 5). Cell numbers increased significantly from sandy to sandy-mud and to mud sediments. This general inverse relationship certainly reflects both the greater surface area and the higher organic matter content in fine sediments as compared to coarse ones (Dale, 1974 ; Tanoue and Handa, 1979). However, the question still remains open whether the bacteria simply respond to higher organic matter level originally present in fine sediments or whether higher bacterial numbers in fine sediments results in higher organic matter accumulations (DeFlaun and Mayer, 1983).

Relationship between biomass and organic material

Relationships between bacterial biomass and organic matter content of the sediments seem to be very complex. In sediments of the Kiel Bight, bacterial carbon increased significantly from sandy beach sediments to sandy-mud sediments. However, in muddy sediments bacterial biomass did not further increase despite of the more than twofold increase in organic matter as compared to sandy-mud sediments (Fig. 6).

On an average, bacterial carbon in sediments of the Kiel Bight accounted for 0.7 ± 0.2% (standard deviation of the mean ; 76 observations) of the sediment organic carbon. In the literature, higher values have been reported (1.2 %, Dale, 1974 ; less than 2%,

Cammen, 1982) which may be partly due to methodological differences in calculating bacterial biomass. Bacterial carbon as percentage of the total sediment carbon varied dependent upon the types of sediments in the Kiel Bight. The lowest percentage was found in sandy beaches (0.5 ± 0.1% ; 36 observations), a considerable higher percentage in sandy-mud sediments (0.9 ± 0.2% ; 21 observations), and again a lower percentage in muddy sediments (0.7 ± 0.2% ; 19 observations). Parallel, the average bacterial cell weight varied in the different types of sediment investigated in the Kiel Bight. For sandy sediments an average cell weight of $15.8 \times 10^{-9} \pm 2.5 \times 10^{-9}$ µg of bacterial carbon was calculated as compared to $20.4 \times 10^{-9} \pm 3.9 \times 10^{-9}$ µg (sandy-mud sediments) and $11.3 \times 10^{-9} \pm 1.7 \times 10^{-9}$ µg of bacterial carbon (muddy sediments).

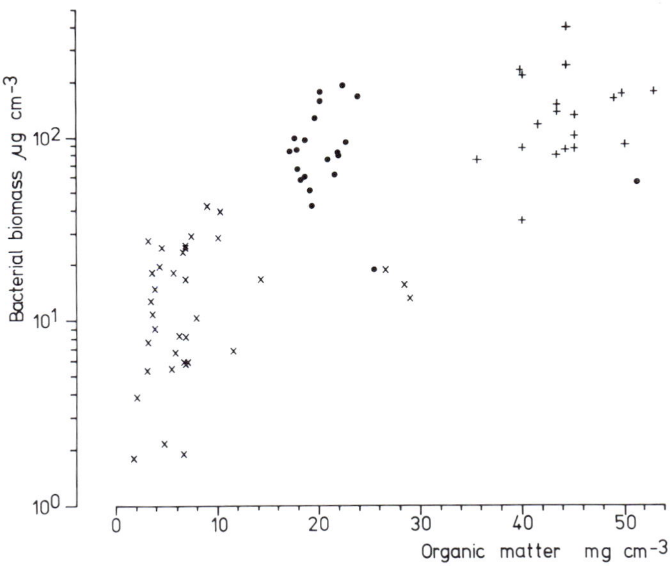

Fig. 6. Bacterial biomass versus total organic matter content on a hypothetical horizontal profile from sandy beaches (x) to sandy-mud sediments (o) and to muddy sediments (+) of the Kiel Bight (Baltic Sea, FRG).

As an explanation for the observations described above, both the concentration and the suitability of the sediment organic material have to be considered. In the Kiel Bight, the sandy-mud sediments obviously offer optimal conditions for the development of the bacterial populations, sustaining the highest total bacterial biomass, the highest average cell weight and the highest percentage of bacterial carbon as compared to the total sediment carbon. In sandy beach sediments on the one hand and in muddy sediments on the other hand, conditions for the development of bacterial populations have to be regarded as suboptimal. In sandy sediments the concentration of organic material may be the limiting factor for bacterial growth. In muddy sediments, however, much of the organic material may consist of highly refractory compounds thus restricting bacterial biomass development.

Size spectrum of biomass

Grouping the bacteria into different size classes revealed a characteristic spectrum of bacterial biomass in sediments of the Kiel Bight. Generally, small-size bacteria (volume < 0.3 μm^3) accounted for the major part of the total bacterial biomass, followed by medium-size bacteria (volume 0.3-0.6 μm^3) and large-size bacteria (volume > 0.6 μm^3). This pattern fits into the average "Sheldon" spectrum, a characteristic distribution of benthic biomass with two main peaks in the largest (> 2 mm ; corresponding to macrofauna) and in the smallest size classes (< 2 µm ; corresponding to bacteria, respectively ; cf. Shwinghamer, 1981). Deviations from this typical pattern in the distribution of benthic biomass have been interpreted by the author as an expression of the effect of exogenous disturbance factors. This interpretation can be related to deviations from the "normal" bacterial biomass spectrum as well. The input of the phytoplankton blooms into the sediments in autumn and spring, respectively, as well as mass abundances of predators let to drastic shifts in the size spectrum of bacterial biomass in sediments of the Kiel Bight (cf. below).

SEASONAL DEVELOPMENT OF BACTERIAL COMMUNITIES

In the literature information on the seasonal development of benthic bacteria is limited. From the investigations of Montagna (1982) and Cammen (1982) no general trend for seasonal variations of benthic bacteria could be detected. The studies of Rublee (1982) and DeFlaun and Mayer (1983) revealed a positive correlation between benthic bacteria and temperature. Detailed investigations in a sandy-mud sediment of the Kiel Bight using a high time resolution in sampling demonstrated strong seasonal variations in the development of bacterial populations closely related to specific ecological situations and sedimentation events (Meyer-Reil, 1983, 1984b).

Input of organic material into the sediment

In recent years it could be shown that in boreal marine systems the main sedimentation events and therefore the main food supply for the benthos occur in autumn and spring, respectively. Studies in the Kiel Bight have demonstrated that large amounts of the primary produced material in autumn and spring do not enter the pelagic food web but rather settle onto the sediment. This material already represents 2/3 of the total yearly input from the pelagic into the benthic system (Smetacek, 1980). Investigations of Peinert et al. (1982) and Gral et al. (1983) followed the biomass development of the autumn and spring phytoplankton bloom in the water column and the sedimentation of the primary produced organic material by sediment traps.

In sediments of the Kiel Bight, three periods of accumulation of organic material could be distinguished : in autumn, winter and spring (Fig. 7 ; Meyer-Reil, 1983). The enrichment of organic material during November could be traced back to the breakdown and sedimentation of the autumn phytoplankton bloom composed of dinoflagellates and diatoms. Total organic matter, protein and carbohydrate accumulated in distinct, separate peaks. During winter a slow continuous increase of organic material was observed in the sediment surface. The organic matter consisted of resuspended sediment, material from terrestrial origin as well as of macrophyte debries eroded by winter storms. The breakdown of the spring phytoplankton bloom (mainly diatoms) let to an enrichment of organic material in the sediment surface during late March to mid April. Again separate peaks were recorded for total organic matter, protein and carbohydrae. These accumulation periods termed "autumn", "winter"- and "spring-input" turned out to be of high relevance for the development of the benthic communities in sediments of the Kiel Bight (Gral et al., 1983 ; Meyer-Reil, 1983, 1984b).

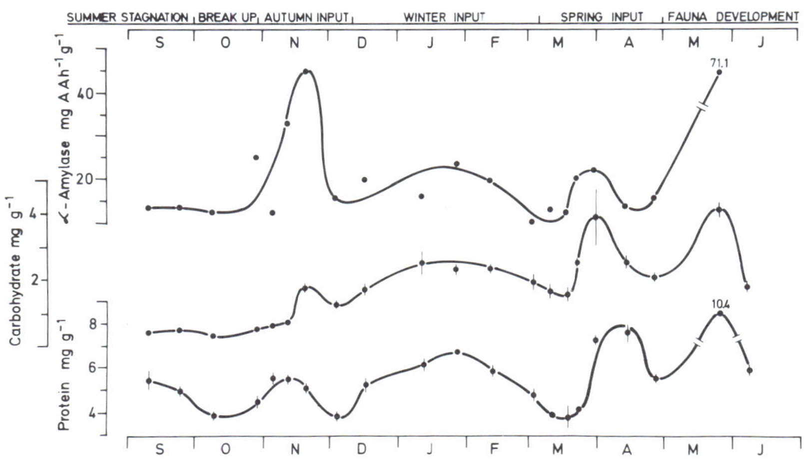

Fig. 7. Seasonal variations in concentrations and exoenzymatic decomposition rates of organic material in the O to 1 cm horizon of a sandy-mud sediment (water depth 18 m) of the Kiel Bight (Baltic Sea, FRG). Illustrated are : concentrations of protein and carbohydrate, and activity of α-amylase (mg of amylopectin azure decomposed per g of dry weight sediment per hour). The headline on top characterizes specific ecological situations and events affecting the sediments.

Heat production as measurement of community metabolism.

Direct calorimetry (heat production) represents a reliable method for the estimation of total benthic community metabolism, because it is a direct measurement of the energy flow through the system. Although the heat release from the activity of extracellular enzymes and from chemical oxidations are also included, the latter two components are thought to be of minor importance for total heat loss from sediments (Pamatmat, 1982).

As an immediate response to the "autumn"- and "spring-input" in sediments of the Kiel Bight, heat production culminated. A less obvious response was found during "winter-input". However, considering the temperature dependence of benthic metabolism, there was still a considerable and long lasting heat production during this period (Graf et al., 1983). Whereas heat production comprises all types of benthic metabolism, electron-transport-activity (ETS) relates to the activity of respiratory chains. As pointed out by Pamatmat et al. (1981), the quotient between heat production and ETS-activity should serve as a qualitative indicator for changes in the type of metabolism. Following the "autumn-" and "spring-input", a strong increase in this quotient was observed, demonstrating a shift in the type of benthic metabolism towards fermentation. This coincided with suboxic conditions in the sediment surface mainly caused by biological oxygen consumption (Graf et al., 1983).

Decomposition of particulate organic material.

The main portion of the input of organic material into the sediment is particulate organic carbon which has to be enzymatically decomposed, at least partly, prior to incorporation into cells. In this process, extracellular enzymes are involved which are secreted by living cells (Corpe and Winters, 1972) or liberated during the lysis of dead and decaying cells. As shown by Burns (1980) some of these enzymes may retain their activity outside the cells by the formation of humic-enzyme complexes bound to clay particles.

The accumulation of organic material in sediments of the Kiel Bight during "autumn-", "winter-" and "spring-input" let to corresponding stimulations in the enzymatic decomposition rates of carbohydrate and protein (activity of α -amylase, proteolytic enzymes, Fig. 7). Exoenzymatic responses were highest in autumn as compared to winter and spring. This is obviously a reflection of both the higher temperature and the higher benthic biomass in autumn. During "autumn-" and "spring-input" a stimulation of enzymatic decomposition rates already occurred when concentrations of particulate organic material started to accumulate in the sediment surface indicating an induction of enzymatic activities by increasing concentrations of suitable substrates (Meyer-Reil, 1983). There is strong evidence from laboratory and field data that under anoxic conditions the enzymatic decomposition of protein is retarded. During summer stagnation, an anoxic period in which hydrogen sulfide is prevailing in sediments of the Kiel Bight, protein accumulated. The concentration of the stored protein was even comparable to that measured during "autumn"- and "spring-input". Following the introduction of oxygen into the sediment (break up of summer stagnation), protein concentrations significantly decreased. Parallel peaks in heat production and ATP imply that the stored protein was rapidly consumed and incorporated into benthic biomass (cf. Graf et al., 1983 ; Meyer-Reil, 1983).

Bacterial development in autumn and spring.

As shown for sediments of the Kiel Bight, bacteria reacted to the "autumn-" and "spring-input", respectively, with two separate peaks. The first peak already occurred when concentrations of organic material started to accumulate in the sediment surface. This demonstrates that bacteria almost immediately responded to the availability of decomposable organic material. The second peak in bacterial parameters coincided with the main input of organic material following the final breakdown and sedimentation of the phytoplankton blooms (Fig. 8).

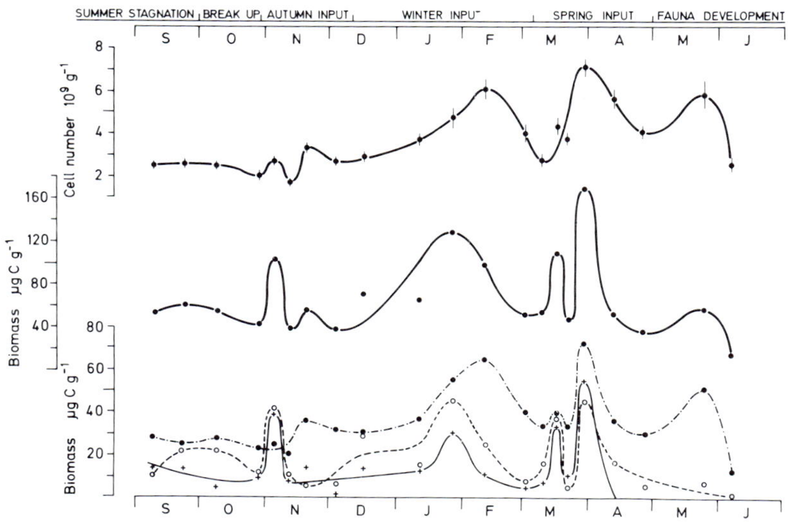

Fig. 8. Seasonal variations in microbiological parameters in the 0 to 1 cm horizon of a sandy-mud sediment (water depth 18m) of the Kiel Bight (Baltic Sea, FRG). Illustrated are: total number of bacteria, total bacteria biomass and biomass spectrum. The biomass spectrum comprises : small-size bacteria (volume < 0.3 m^3 ; closed circles), medium-size bacteria (volume 0.3-0.6 μm^3 ; open circles), and large-size bacteria (volume > 0.6 μm^3 ; crosses). For the headline on top cf. Fig. 7.

The bacterial population faced with the "autumn-input" was derived from an anoxic population (fermentative bacteria, sulfate reducers) which prevailed during summer stagnation. Within this population the input of freshly produced organic material caused a drastic shift. Bacteria primarily reacted with a strong increase in cell volume (biomass production). Deviating from its "normal" distribution (cf. above), the size spectrum was dominated by medium and large-size cells. Following the final breakdown and sedimentation of the autumn phytoplankton bloom, the bacteria subsequently responded with cell division (increase in cell number).

Compared to autumn, the history of the bacterial population faced with the "spring-input" was quite different. During winter oxic conditions prevailed in the sediment. Due to erosion of the sediment caused by winter storms, the bacterial population was declining. The "spring-input" hit an impoverished bacterial community which immediately reacted with a strong increase in cell volume, but only a small increase in cell number. Again, deviations from the normal size spectrum were observed : small, medium and large-size cells almost equally contributed to the total biomass. Following the final breakdown and sedimentation of the spring phytoplankton bloom, bacterial number and biomass responded with a second stimulation.

Bacterial development in winter.

During winter resuspended sediment, terrestrial material and eroded macrophytes represent an additional food supply for benthic bacterial populations of the Kiel Bight. Bacterial number, biomass and cell-division activity showed a slow continuous increase up to values that were even higher as compared to those obtained following the "autumn-input" (Fig. 8). Taking into account the more refractory nature of the organic material and the low temperature with reduced matabolic activity rates, the accumulation of bacterial biomass during winter is surprising (cf. temperature-dependent development of benthic bacteria reported by DeFlaun and Mayer, 1983). However, the relatively long time the bacterial population had available for its "undisturbed development" and the limited number of grazers have to be considered. With regard to the nature of the organic material and the slow continuous development, the response of the bacterial community to the "winter-input" differed basically from the spontaneous bacterial development following the input of the phytoplankton blooms in autumn and spring, respectively.

Bacteria and fauna development.

The development of the benthic fauna in spring greatly influenced the composition of the bacterial community in sediments of the Kiel Bight. Through the activity of polychaetes the sediment surface was firmely glued together. This ecological situation was reflected by a bacterial population which consisted of a high number of almost exclusively small-size cells with a corresponding low biomass (Fig. 8). Since the bacteria actively grew as demonstrated by a high number of dividing cells, nutrient deficiency could not be the reason for the impoverishment of the bacterial population. More likely, preferential grazing on medium and large-size bacteria let to the restriction in the size spectrum of the bacteria. Although literature regarding interrelationships between bacteria and fauna components is sparse, there is some evidence for the stimulation of bacterial activity by grazing (Gerlach, 1978 ; Morrison and White, 1980).

DIURNAL FLUCTUATIONS OF BACTERIAL POPULATIONS

The strong diurnal rhythms of benthic primary production (Jorgensen et al., 1979 ; Karg, 1979 ; Revsbech et al., 1981) imply a coupling between autotrophic and heterotrophic processes in shallow water sediments. Evidence for the existence of diurnal rhythms in benthic bacterial activities obviously closely related to primary production was obtained from investigations of sediments below sea-grass beds in Moreton Bay, Queensland, Australia (Moriarty and Pollard, 1982) and from studies in sandy sediments of the Kiel Bight, Baltic Sea, FRG (Meyer-Reil and Graf, unpublished data). Since the latter study comprises various parameters related to benthic biomass and activity, the discussion of diurnal fluctuations of bacterial population will be based upon these results.

Benthic biomass and activities.

In sandy sediments of the Kiel Bight sampled during a 36 hour cycle in summer, strong diurnal fluctuations occurred in total organic matter, benthic biomass and activities (Fig. 9, 10). Total organic matter and living benthic biomass (derived from ATP measurements) accumulated around midnight and decreased during the day.

Bacterial biomass, however, revealed an opposite fluctuation pattern : maxima were observed in late afternoon and minima between midnight and early morning (cf. Fig. 9).

As indicated by the almost constant ratio between total organic matter and protein, the composition of organic material did not change during the day/night cycle. Assuming a conversion factor of 200 for calculating organic carbon from ATP, and assuming that carbon represents 50% of the total organic material, living biomass on an average accounted for 5% of the total organic carbon. Bacterial biomass made up 5% of the total living biomass.

Benthic activities were well correlated among each other despite of the fact that quite different parameters of heterotrophic activity were measured such as total metabolic activity (heat production), exoenzymatic decomposition of particulate organic material (activity of α-amylase), and bacterial uptake of dissolved organic substrates (^{14}C-labelled glucose). Benthic activities increased during the morning, culminated around noon and decreased during afternoon and night (cf. Fig. 10). A corresponding diurnal pattern in bacterial activity was reported by Moriarty and Pollard (1982) for sediments below seagrass beds based upon the incorporation of thymidine into DNA.

Trophic interrelationships.

At least in periods with sufficient light supply, shallow water sediments may represent self-supporting systems governed by benthic primary production. It may be speculated that heterotrophic benthic activities are stimulated by the excretion of substances from the benthic primary production (microphytobenthos) which is initiated by light in the early morning. Bacteria take up the excretion products and respond with a subsequent increase in biomass (note the time lag between bacterial uptake of substances and increase in biomass, Fig. 9, 10). Bacteria as basic members of the food chain are grazed by meio- and macrofauna. As a consequence, total benthic activity (heat production) and enzymatic decomposition of particulate organic material in the sediment as well as in selected meiofauna organisms (cf. Faubel and Meyer-Reil, 1983) increase. After maximum values are reached around noon, benthic activities decline towards the afternoon and night obviously due to a decreasing supply with primary produced material.

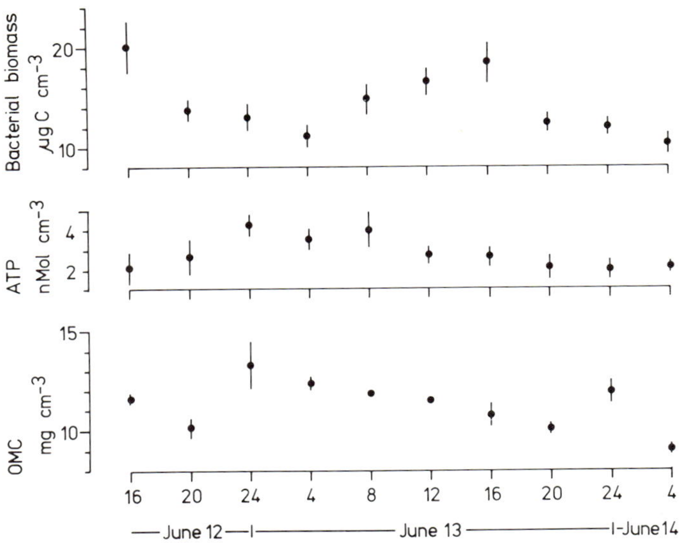

Fig. 9. Diurnal fluctuations of total organic matter, ATP, and bacterial biomass in the 0 to 1 cm horizon of a sandy sediment (water depth 10 m) of the Kiel Bight (Baltic Sea, FRG) sampled every 4 hours during a 36 hour cycle (June 12 to June 14, 1980).

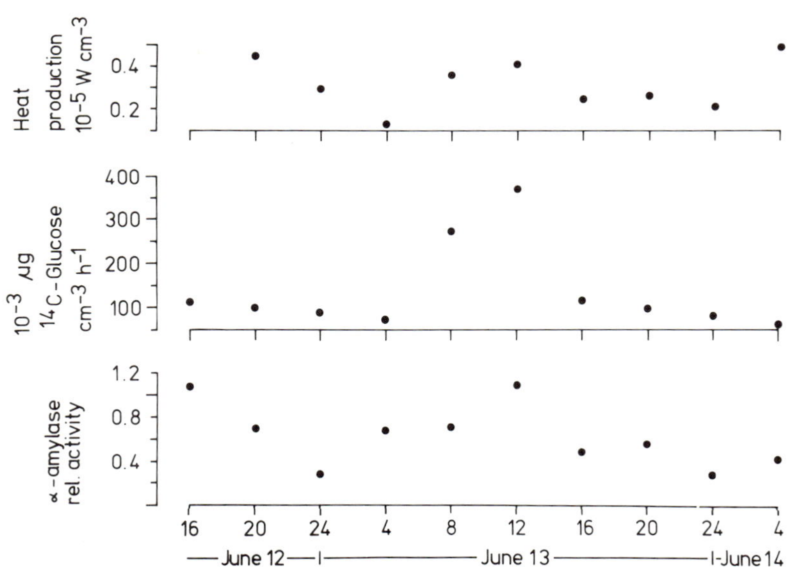

Fig. 10. Diurnal fluctuations of exoenzymatic decomposition rates of carbohydrate (activity of α-amylase), bacterial uptake of ^{14}C-glucose, and heat production in the 0 to 1 cm horizon of a sandy sediment (water depth 10 m) of the Kiel Bight (Baltic Sea, FRG) sampled every 4 hours during a 36 hour cycle (June 12 to June 14, 1980).

ESTIMATES OF BACTERIAL PRODUCTION

Since bacteria play an important role as basic members of the food chain in sediments (Yingst and Rhoads, 1980 ; Gerlach, 1978), information on bactrial production is urgently needed. In the literature, however, reliable data on bacterial production are sparse. Certainly this is a reflection of the methodological problems involved. Different approaches have been applied to gain information on bacterial production in sediments (cf. Table 1).

Glucose uptake.

In sandy beaches of the Kiel Bight, actual uptake (flux) of glucose by bacteria amounted to 0.1 µg of carbon (glucose) per gram of dry weight sediment per h (summer conditions, average of 12 observations ; Meyer-Reil, 1978 ; Meyer-Reil et al. 1980). In these sediments approximately 2% of the total dissolved organic carbon (DOC) existed in the form of labile carbon (1% amino acids, 1% monosaccharides) from which glucose roughly represents one quarter (Meyer-Reil et al., 1978). Under the assumption that bacteria take up the remaining three quarters of the DOC with the same velocity, total bacterial carbon uptake would amount to 0.4 µg per g per h. Taking into account respiration (40% as an average of the respiration of different organic substrates) and excretion (estimate of 10%, no data available) bacteria would produced 0.2 µg of carbon per g per h, which is equivalent to 28 mg of carbon per m^2 of sediment per day (Table 1).

TABLE 1.

Summary of data concerning bacterial production in sediments of different areas.

Sediment	Method	Production (mg C m^{-2} d^{-1})
(1) Sandy beaches, Kiel Bight, Baltic Sea, FRG	Glucose uptake	28
(2) Sediment associated with seagrass beds Moreton Bay, Queensland, Australia	Thymidine incorporation	12
(3) Sand, Nearshore western Atlantic Ocean Sapelo Island, Georgia, USA	Thymidine incorporation	100-800
(4) Sandy-mud/mud Kiel Bight, Baltic Sea, FRG	Seasonal changes in biomass	
Autumn		140/370
Winter		20/10
Spring		300/120
(5) Sand, Kiel Bight, Baltic Sea, FRG	Diurnal changes in biomass	80

Authors : (1), Meyer-Reil et al. (1980) ; (2), Moriarty & Pollard (1982) ; (3), Fallon et al. (1983) ; (4), Meyer-Reil (1983, and unpublished) ; (5), Meyer-Reil (unpublished).

This means that between 10 and 30% of the microphytobenthos primary production in this area (Karg, 1979) is fixed by bacterial secondary production, a range which sounds reasonable adding at least some confidence to the bacterial production data calculated. However, as pointed out above, various assumptions have to be considered.

Thymidine incorporation.

The incorporation of thymidine into DNA was used by Moriarty and Pollard (1982) and Fallon et al. (1983) to calculate benthic bacterial production. For sediments below seagrass beds in Moreton Bay, Queensland, Australia, the former authors reported a bacterial production of 12 mg of carbon per m^2 of sediment per day. Nearshore sediments from the western Atlantic Ocean, Sapelo Island, Georgia, USA, revealed a much higher bacterial production (100-800 mg of carbon per m^2 per day ; cf. latter authors). The lower limit of these data well agree with the bacterial production extrapolated from seasonal and diurnal changes in bacterial biomass in sediments of the Kiel Bight (cf. below, cf. Table 1). However, calculating bacterial production from the incorporation of thymidine requires a number of assumptions which are difficult to verify. Among these, the question of isotope dilution and the validity of conversion factors to calculate production from bacterial uptake of thymidine seem to be most important (for a detailed discussion cf. papers mentioned above and literature cited therein).

Diurnal and seasonal changes in biomass.

Calculating bacterial production from changes in bacterial biomass in samples collected frequently during short periods of time (range of hours, cf. Fig. 9) at a fixed location primises to offer the most reliable estimates. By this method the addition of exogenous substrates and artificial conditions during the incubation in the laboratory can be avoided. Except for the conversion from bacterial volume to weight in terms of carbon (cf. above), no conversion factors or assumptions have to be applied. However, other problems arise. A sufficient number of subsamples have to be collected by divers at each time interval from a defined sediment area to avoid artefacts due to patchiness. Since bacterial excretion of substrates and grazing on bacteria between the sample intervals can not be accounted for, the values gained certainly present an underestimation of bacterial production which is difficult to evaluate.

Base upon studies of diurnal fluctuations at a shallow water sediment in the Kiel Bight (cf. above), a bacterial net production of 80 mg of carbon per m^2 per day was calculated. Further estimates of bacterial production were derived from investigations of seasonal varaitions in bacterial biomass at two sediment stations in deeper waters of the Kiel Bight. As response to the input of the phytoplankton bloom in autumn, bacterial production amounted to 140 and 370 mg of carbon per m^2 per day (sandy-mud and muddy sediment, respectively). The corresponding values in spring were 300 and 120 mg of carbon per m^2 per day, respectively (cf. Table 1). As it was pointed out above, these values well agree with the lower range of bacterial production data derived from the incorporation of thymidine into DNA (Fallon et al., 1983).

CONCLUSION

From the information available it can be accepted that bacteria colonize sediments in high number and biomass. Although they represent less than 1% of the total sediment organic carbon, bacteria contribute significantly to the benthic biomass thus stressing their important role as nutrient source for the benthic fauna. Between the distribution of bacteria and sediment properties such as grain size and organic matter, close correlations exist. However, as shown for the relationship between organic material and bacterial biomass, interpretations are difficult to derive.

Processes within the benthic bacterial community may occur in very short time scales. The strong diurnal rhythms observed indicate a close coupling between autotrophic and heterotrophic processes in shallow water surface sediments. In boreal marine ecosystems the seasonal development of the benthic bacterial populations turned out to be strongly influenced by certain ecological situations and events in the sediment, from which the input of the phytoplankton blooms in autumn and spring, respectively, the accumulation of organic material during winter, and the development of the benthic fauna in spring were most important. The enrichment of organic material in the sediment surface let to corresponding stimulations in the enzymatic decomposition rates of particulate organic material. Bacteria immediately reacted on the availability of organic material with a drastic shift in the size distribution of biomass.

Reliable data on benthic bacterial production are urgently needed. From the very few data based upon quite different approaches, production ranges between 10 and some 100 mg of bacterial carbon per m^2 of sediment per day. These estimates already illustrate the importance of bacteria in the turnover of organic material in sediments.

Although basic information on the spatial and temporal distribution of benthic bacterial biomass and activity is available, a number of questions still remain open. Among these, the measurement of bacterial production, seasonal variations in the metabolic activity of bacterial populations, and interactions between bacteria and the benthic fauna need further attention.

ACKNOWLEDGMENTS

This work was supported by the German Research Council (Sonderforschungsbereich 95 der Deutschen Forschungsgemeinschaft). I would like to thank all of my colleagues espacially Dr. G. Graf for supplying me with their data and their helpful comments.

REFERENCES

Bakken, L.R. and Olsen, R.A., 1983. Buoyant densities and dry-matter contents of microorganisms : conversion of a measured biovolume into biomass. Appl. Environ. Microbiol., 45: 1188-1195.

Burns, R.G., 1980. Microbial adhesion to soil surfaces: consequences for growth and enzyme activities. In: R.C.W. Berkeley, J.M. Lynch, J. Melling, R.P. Rutter and B. Vincent (eds), Microbial adhesion to surfaces. ELlis Horwood Limited, Chichester, pp. 249-262.

Cammen, L.M., 1982. Effects of particles size on organic matter content and microbial abundance within four marine sediments. Mar. Ecol. Prog. Ser., 9: 273-280.

Clough, K.S. and Sutton, J.C., 1978. Direct observation of fungal aggregates in sand dune soil. Can. J. Microbiol., 24: 333-335.

Corpe, W.A. and Winters, H., 1972. Hydrolytic enzymes of some periphytic marine bacteria. Can. J. Microbiol., 18: 1483-1490.

Dale, N.G., 1974. Bacteria in intertidal sediments: factors related to their distribution. Limnol. Oceanogr., 19: 509-518.

DeFlaun, M.F. and Mayer, L.M., 1983. Relationships between bacteria and grain surfaces in intertidal sediments. Limnol. Oceanogr., 28: 873-881.

Ellery, W.N. and Schleyer, M.H., 1984. Comparison of homogenization and ultrasonication as techniques in extracting attached sedimentary bacteria. Mar. Ecol. Prog. Ser., 15: 247-250.

Fallon, R.D., Newell, S.Y. and Hopkinson, C.S., 1983. Bacterial production in marine sediments: will cell-specific measures agree with whole-system metabolism. Mar. Ecol. Prog. Ser., 11: 119-127.

Faubel, A. and Meyer-Reil, L.-A., 1983. Measurement of enzymatic activity of meiobenthic organisms: methodology and ecological application. Cahiers de Biologie Marine, 24: 35-49.

Ferguson, R.L. and Rublee, P., 1976. Contribution of bacteria to standing crop of coastal plankton. Limnol. Oceanogr., 21: 141-145.

Gerlach, S., 1978. Food-chain relationships in subtidal silty sand marine sediments and the role of meiofauna in stimulating bacterial productivity. Oecologia (Berl.), 33: 55-69.

Griffiths, R.P., Hayasaka, S.S., McNamara, T.M. and Morita, R.Y., 1978. Relative microbial activity and bacterial concentrations in water and sediment samples taken in the Beaufort Sea. Can. J. Microbiol., 24: 1217-1226.

Graf, G., Bengtsson, W., Diesner, U., Schulz, R. and Theede, H., 1982. Benthic response to sedimentation of a spring phytoplankton bloom : process and budget. Mar. Biol., 67: 201-208.

Graf, G., Schulz, R., Peinert, R. and Meyer-Reil, L.-A., 1983. Benthic response to sedimentation events during autumn to spring at a shallow-water station in the Western Kiel Bight. I. Analysis of processes on a community level. Mar. Biol., 77: 235-246.

Hargrave, B.T., 1972. Aerobic decomposition of sediment and detritus as a function of particle surface area and organic content. Limnol. Oceanogr., 17: 583-596.

Jones, J.G., 1980. Some differences in the microbiology of profundal and littoral lake sediments. J. Gen. Microbiol., 117: 285-292.

Jørgensen, B.B., Revsbech, N.P., Blackburn, T.H. and Cohen, Y., 1979. Diurnal cycle of oxygen and sulfide microgradients and microbial photosynthesis in a cyanobacterial mat sediment. Appl. Environ. Microbiol., 38: 46-58.

Karg, S., 1979. Vergleichende Produktivitätsuntersuchungen am Mikrophytobenthos in der Kieler Förde. Reports Sonderforschungsbereich, 95, Universität Kiel.

Kepkay, P.E., Cooke, R.C. and Novitsky, J.A., 1979. Microbial autotrophy : a primary source of organic carbon in marine sediments. Science, 204: 68-69.

Krambeck, C., Krambeck, H.-J. and Overbeck, J., 1981. Microcomputer assisted biomass determination of plankton bacteria on scanning electron micrographs. Appl. Environ. Microbiol., 42: 142-149.

Meyer-Reil, L.-A., 1978. Uptake of glucose by bacteria in the sediment. Mar. Biol., 44: 293-298.

Meyer-Reil, L.-A., 1981. Enzymatic decomposition of proteins and carbohydrates in marine sediments: methodology and field observations during spring. Kieler Meeresforsch., Sonderh., 5: 311-317.

Meyer-Reil, L.-A., 1983. Benthic response to sedimentation events during autumn to spring at a shallow-water station in the Western Kiel Bight II. Analysis of benthic bacterial populations. Mar. Biol., 77: 247-256.

Meyer-Reil, L.-A., 1984a. Bacterial biomass and heterotrophic activity in sediments and overlying waters. In: J.E. Hobbie and P.J. LeB. Williams (eds), Heterotrophic activity in the sea. Plenum Press, New York, London, pp. 523-546.

Meyer-Reil, L.-A., 1984b. Seasonal variations in bacterial biomass and decomposition of particulate organic material in marine sediments. Arch. Hydrobiol. Beih. Ergebn. Limnol., 19: 201-206.

Meyer-Reil, L.-A., Bölter, M., Dawson, R., Liebezeit, G., Szwerinski, H. and Wolter, K., 1980. Interrelationships between microiological and chemical parameters of sandy beach sediments, a summer aspect. Appl. Environ. Microbiol., 39: 797-802.

Meyer-Reil, L.-A., Dawson, R., Liebezeit, G. and Tiedge, H., 1978. Fluctuations and interactions of bacterial activity in sandy beach sediments and overlying waters. Mar. Biol., 48: 161-171.

Meyer-Reil, L.-A., Schramm, W. and Wefer, G., 1981. Microbiology of a tropical coral reef system (Mactan ; Philippines). Kieler Meeresforsch., Sonderh., 5: 431-432.

Montagna, P.A., 1982. Sampling design and enumeration statistics for bacteria extracted from marine sediments. Appl. Environ. Microbiol., 43: 1366-1372.

Moriarty, D.J.W., 1980. Measurement of bacterial biomass in sandy sediments. In: P.A. Trudinger, M.R. Walter and B.J. Ralph (eds.), Biogeochemistry of ancient and modern environments. Australian Academy of Science, Canberra, and Springer-Verlag, Berlin, pp. 131-139.

Moriarty, D.J.W. and Hayward, A.C., 1982. Ultrastructure of bacteria and the proportion of gram-negative bacteria in marine sediments. Microb. Ecol., 8: 1-14.

Moriarty, D.J.W. and Pollard, P.C., 1982. Diel variation of bacterial productivity in seagrass (Zostera capricorni) beds measured by rate of thymidine incorporation into

DNA. Mar. Biol., 72: 165-173.

Pamatmat, M.M., Graf, G., Bengtsson, W. and Novak, C.S., 1981. Heat production, ATP concentration and electron transport activity of marine sediments. Mar. Ecol. Prog. Ser., 4: 135-143.

Pamatmat, M.M., 1982. Heat production by sediment: ecological significance. Science, N.Y., 215: 395-397.

Peinert, R., Saure, A., Stegmann, P., Stienen, C., Haardt, H. and Smetacek, V., 1982. Dynamics of primary production and sedimentation in a coastal ecosystem. Neth. J. Sea. Res., 16: 276-289.

Revsbech, N.P. and Jørgensen, B.B., 1981. Primary production of microalgae in sediments measured by oxygen microprofile, $H^{14}CO_3^-$ fixation, and oxygen exchange methods. Limnol. Oceanogr., 26: 717-730.

Rublee, P.A., 1982. Seasonal distribution of bacteria in salt marsh sediments in North Carolina. Estuarine Coastal Shelf Sci., 15: 67-74.

Rublee, P.A. and Dornseif, B.E., 1978. Direct counts of bacteria in the sediments of a North Carolina salt marsh. Estuaries, 1: 188-191.

Schwinghamer, P., 1981. Characteristic size distributions of integral benthic communities. Can. J. Fish. Aquat. Sci., 38: 1255-1263.

Smetacek, V., 1980. Annual cycle of sedimentation in relation to plankton ecology in western Kiel Bight. Ophelia, 1, Suppl. 65-76.

Tanoue, E. and Handa, N., 1979. Differential absorption of organic matter by various sized sediment particles in recent sediment from the Bering Sea. J. Oceangr. Soc. Jap., 35: 199-208.

Weise, W. and Rheinheimer, G., 1978. Scanning electron microscopy and epifluorescence investigation of bacterial colonization of marine sand sediments. Microb. Ecol., 4: 175-188.

Weise, W. and Rheinheimer, G., 1979. Fluoreszenzmikroskopische Untersuchungen über die Bakterienbesiedlung mariner Sandsedimente. Botanica Marina, 22: 99-106.

Yingst, J.Y. and Rhoads, D.C., 1980. The role of bioturbation in the enhancement of bacterial growth rates in marine sediments. In: K.R. Tenore and B.C. Coull (eds), Marine benthic dynamics. University of South Carolina Press, Columbia, pp. 407-421.

HEAT PRODUCTION OF MICROORGANISMS IN EUTROPHIED ESTUARINE SYSTEMS - AN EXPERIMENTAL STUDY.

P. LASSERRE[1], T. TOURNIE[1], M. BIANCHI[2] and S. CHAMROUX[1].
[1]Station Biologique de Roscoff, Université de Paris VI & CNRS, 29211 Roscoff
[2]Laboratoire de Microbiologie Marine, CNRS, Univ. de Provence, 13331 Marseille (France)

ABSTRACT

Lasserre, P., Tournié, T., Bianchi, M. and Chamroux, S., 1986. Heat production of microorganisms in eutrophied estuarine systems - An experimental study. In: P. Lasserre and J.M. Martin (eds), Biogeochemical Processes at the Land-Sea Boundary. Elsevier, Amsterdam.

The microbially-mediated degradation of organic materials in estuaries and lagoons exerts important influences on the chemical composition of sediments and waters. In these shallow water systems, the biogeochemical processes occuring at the sediment-water interface are relatively well-known, but rates and mechanisms are ill defined. The temporal behaviour of microorganisms colonising the water-sediment interface from different estuarine and lagoonal environments was studied, in microcosms, to compare their relative ability to recover from experimental eutrophication.

The microcalorimetric data and simultaneous studies of the microbial physiological potentialities indicate that the power-time curves describe dissipative structure, with reproducible patterns in the form of temporal successions of microorganisms. The evolution of the specific rate of heat dissipation ($\dot{q}$) showed differences in relation to seasonal thermal regimes. A characteristic feature is the increase in the specific heat production rate, shortly after nitrogen enrichment of the microcoms, followed by a decrease indicating subsequent adaptational changes in populations. These results offer an example of the value of direct microcalorimetry in studying homeostatic capabilities of ecosystems. Here, heat dissipation is an adequate parameter of the ability of the system to return, after transitory oscillations, to a new steady state (a parameter of "resilience").

INTRODUCTION

Biogeochemical processes in organic-rich estuaries and lagoons are dominated by the influence of microbially mediated degradation of recently deposited organic matter. These processes are relatively well defined and they take place notably at the sediment-water interface (Martens, 1982 ; Nixon, 1984). Nevertheless, the difficulty in providing a characterization of the dynamics of microbiological processes induced by eutrophication still remains. The kinetics of biogeochemical transformations, brought about during the degradation of complex mixtures of material of natural and anthropogenic origin, by ill defined mixed populations of microbes such as occurs in the marine coastal systems, are not easily studied by the reductionist techniques applicable to homogeneous systems.

Several attempts have been made to describe the characteristics of the aerobic and anaerobic parts of the metabolism of sediments, using ampoule calorimeters (Pamatmat, 1984). Furthermore, flow microcalorimeters proved to be a suitable method for quantifying rapid (hours) changes occuring in micoorganisms metabolism occuring in shallow water eutrophied systems located in three contrasting localities : the Arcachon basin, the Gironde estuary (Southwestern Atlantic coast) and the "Aber" (ria) of Roscoff in north Brittany (Lasserre, 1980, 1984 ; Lasserre and Tournié 1984 ; Tournié and Lasserre, 1984). Results are reported here which provide an indication of the temporal behaviour of complex mixed populations of microorganisms at the water-sediment interface. An understanding of such dynamic processes make a significant contribution to improving our predictions of responses of systems at the land-sea boundary to eutrophication resulting from human activities as well as natural phenomena.

METABOLISM AND HEAT PRODUCTION

Transformation of matter during metabolism by living organisms is accompanied by energy change. Some chemical energy is conserved, for short or long periods, but there is always a net loss of energy manifested as heat loss whether organic matter is completely mineralized or not. The heat flow from living organisms, regardless of metabolism type is a direct result of metabolism. Calorimetry is a sensitive indicator of the energy changes in biological systems, giving information about the rates and extent of reactions, no matter how complex the processes occurring. Since any chemical or physical process will be accompanied by an enthalpy change, the method is completely general. It might seem that there would be considerable difficulty in interpreting results, since resolution of the processes taking place would become very complex in living organisms. However, the practical situation is not usually so complex as it appears at first sight.

In biological systems, the reaction can be considered to take place in solution both at constant pressure and constant volume, so that thermodynamic differences of definition between energy and enthalpy can be disregarded and the change in energy content accompanying a reaction considered to be the change in enthalpy or heat of reaction, ΔH. This quantity, then corresponds to the experimentally measured heat evolution, Q, during a reaction in an isolated system in which no external work is done, with the convention that an exothermic reaction corresponds to a decrease in energy, so that ΔH is negative. The amount of heat evolved is proportional to n the number of moles of reaction which have taken place :

$$Q = - n \, \Delta H \quad (1)$$

or where several simultaneous reactions occur the total heat produced is the sum of the individual processes :

$$Q = - \sum_{0}^{k} n_j \, \Delta H_j \quad (2)$$

(subscripts 0 to k characterize the k internal processes occurring, n_j is the number of moles of reaction j that have occurred since time zero and ΔH_j is its heat of reaction/mole).

Over the last ten years, the technical development of calorimetry has reached such a stage that even very slow processes connected with a low rate of heat output, such as occur in biological systems, may be detected. A wide variety of particular designs have been used in modern biological calorimetry and applied to molecular, cellular, multicellular, and complex systems studies (general reviews by Spink & Wadsö, 1976 ; Lamprecht & Zotin, 1978 ; Beezer, 1980 ; Gnaiger, 1983).

Microcalorimetric characterization of metabolic trends in aquatic systems

Microcalorimetry has been used to demonstrate the microbial activity in normal and acidified soils (Mortensen et al., 1973 ; Ljungholm et al., 1979 a,b). Other workers have reported results of experiments where total heat production in the marine sediment, maintained in closed calorimetric vessels, has been assessed (Pamatmat, 1980,1984). Recently, the heat production of marine bacteria has been studied by Wagensberg et al. (1978), Gordon and Millero (1980).

We have recently developed a microcalorimetric technique for quantifying the energetic changes of microorganisms colonizing a sea-water-sediment interface in experimental microcosms. The inherent low specificity of direct microcalorimetry and, moreover, the high sensitivity and reliability of modern microcalorimeters proved advantageous, since unknown and subtle events, not shown by more specific methods, may be detected.

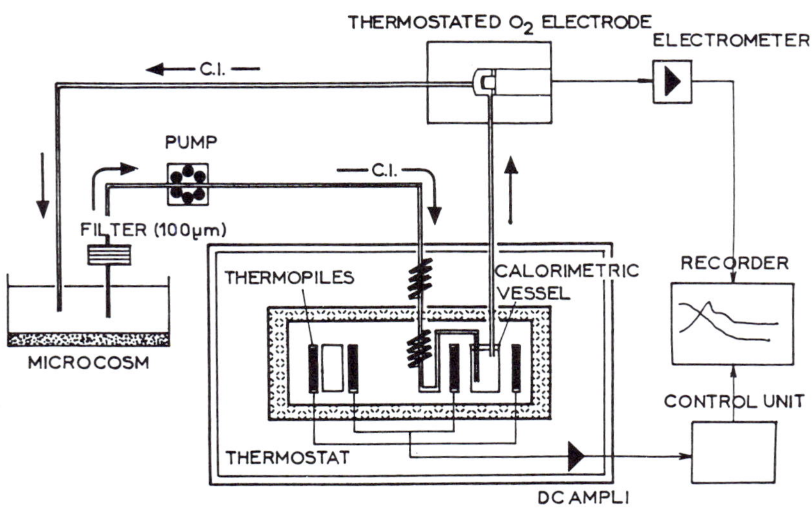

Fig. 1. Diagrammatic view of the microcalorimetric and oximetric circulation system for the study of metabolic activity in experimental microcosms, at the water-sediment interface : CI, circulating interface (for details cf. Lasserre & Tournié, 1984).

It appears, therefore, that the measurement of heat flow using direct calorimetry, is a suitable technique for biogeochemical studies in aquatic systems since under carefully controlled conditions a reproducible "fingerprint" (or power-time curve) is obtained.

Our published results (Lasserre, 1980 ; Lasserre & Tournié, 1984 ; Lasserre, 1984 ; Tournié & Lasserre, 1984), report basic methodological problems involved in the continuous and simultaneous registration of heat production and oxygen pressure changes occuring in a circulating marine interface pumped from just above the sediment. This "circulating interface" is taken through the microcalorimeter (acting as a wattmeter), then to an oxygen electrode and is then returned to the microcosm (Fig. 1).

The technique provides a reliable method by which changes in the metabolic activity at the seawater-sediment interface can be detected and quantified on an empirical level. The underlying metabolic processes have been identified by adding ^{14}C-glucose to the circulating interface and measuring the ATP pool level. The microcosms are experimentally eutrophicated using nitrogen enrichments (bactopeptone or amino acid cocktail). These enrichments (0.5 to 4.6 mmol N L^{-1}) correspond to situations encoutered in estuaries and lagoons subject to excessive organic inputs originating from urban wastes and agricultural runoff.

The power-time curves (PTCs) obtained with microcosms experimentally enriched and composed of undisturbed cores of sediment and water are highly reproducible (Fig. 2).

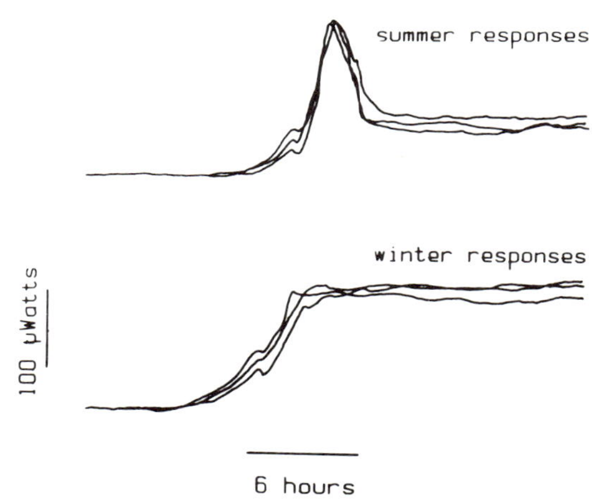

Fig. 2. Power-time curves obtained in replicate experiments after peptone treatment (4mg/ml) : summer response in August, winter response in February.

After nitrogen enrichment, the exponential transitory phase of the power-time curve is clearly correlated with aerobic glucose metabolism, ATP pool level and oxygen tension (pO_2). The microevents noted on PTC during the exponential phase, are also observed on the oxygen-time curve (OTC) and a straight linear correlation is observed between the heat production and O_2 utilisation . Conversely, after the exponential transitory phase, the system shows great differences between batch cultures. The decrease in heat production (more than 100 µW/ml in 2-3 h) occuring after the exponential transitory phase and preceeding the secondary steady state in summer experiments is not correlated with a decrease in ATP, in ^{14}C-particulate matter level and in oxygen uptake rate, as occurs in batch cultures (Belaich, 1980 ; Forrest, 1972). The amplitude of seasonal variations in pO_2, measured at the end of the exponential phase, clearly demonstrates that the reduction of the metabolism, in term of heat production, is not produced by artefact (e.g. oxygen depletion in the tubing) but rather corresponds to a seasonal modification in the integrated metabolic characteristics of the microcosms (Fig. 3).

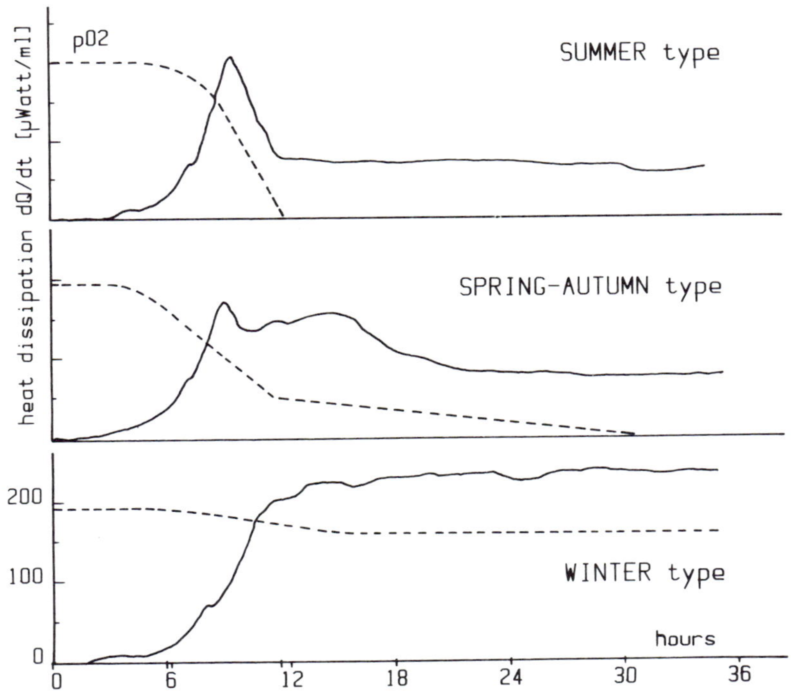

Fig. 3 : Typical seasonal microcalorimetric and oximetric patterns produced in experimental microcosms, at the sea-water - sediment interface, after an experimental eutrophication : A, unimodal summer PTC ; B, bimodal spring-autumn PTC ; C, sigmoïd winter PTC, with their respective oxygen-time curves (OTCs).

Calorimetric and oximetric responses show alternate periods of homogeneity (summer and winter months) and of heterogeneity (spring and autumn months) : periods of homogeneity were characterized by highly reproducible PTCs and OTCs ; periods of heterogeneity displayed transition stages between summer and winter types. The spring heterogeneous PTCs obtained from March to June can be interpreted as a gradual evolution from sigmoid (winter type) to unimodal (summer type) PTCs .

The autumnal transition period, extending from October to mid-January presented, both for PTCs and OTCs, an alternation of the three basic seasonal patterns : summer type, winter type and transition spring type, with increasing frequency of winter type responses as winter approached. Our results suggest that heat productions and their oxidative counterparts are correlated with seasonal trends. It is noteworthy that calorimetric and oximetric measurements have been performed at a constant temperature (18°C ± 1), all year round. Therefore, the difference noted between winter and summer PTCs and OTCs are very probably related to long term cyclic seasonal fluctuations, occuring in the estuarine and lagoonal environments studied. Moreover, our results indicate that seasonally-induced environmental stress (temperature, eutrophication) could determine conservative metabolic responses of the ecosystem, of adaptative nature.

The experimental supply of organic matter, simulating an acute eutrophication in the natural environment, will induce the activation of adaptative processes (corresponding to the transitory phase in the PTCs). The latter could be interpretated in terms of physiological processes (e.g. switch in aerobic/anaerobic metabolisms) or demographic processes (e.g. population increase, succession), leading to a new steady state, and characterized by a constant heat dissipation (1) very high in winter (145 μW), indicating a poorly adapted system, i.e. requiring large energy amounts for maintenance activity, (2) three times less in summer (45-50 μW), indicating a low requirement in energy for maintenance (which may indicate a good adaptation to eutrophic and anoxic conditions). In summer, when no oxygen is left in the microcosm after an enrichment experiment, there is always a stable heat production level of 40 - 50 μW, completely inhibited by formalin, during the experimental time (until 1 week) . Therefore, it is probable that this secondary steady state is due to anaerobic metabolism.

MICROCALORIMETRIC STUDIES AND MICROBIAL DYNAMICS

Although many problems remain to be solved, the results indicate that direct microcalorimetry provides new information on the overall temporal behaviour of complex mixtures of microorganisms colonising the water-sediment interface of coastal environments. The observed microcalorimetric trends were related, as a working hypothesis, to season-dependent adaptation of microorganism populations colonising the sediment-water interface (Tournié & Lasserre, 1984). It was essential to reach a better understanding of the population successions and physiological properties of microorganisms. Recent works, developed on the basis of this hypothesis, with microbiologists in Arcachon (M. Bianchi and coll.) and in Roscoff (S. Chamroux and coll.) showed specific seasonal adaptation of natural marine bacterial populations in response to experimental eutrophication.

Bacteriological studies

Rates of heat production dQ/dt were measured in parallel with oxygen tension as described above. Water samples were taken simultaneously in the microcosms during the course of the (24-hour duration) for bacterial dynamics analysis. For each experiment, water samples were taken over a 24 hour period from control and experimental microcosms. Plate counts (Oppehheimer & ZoBell, 1952) were made in order to determine the viable heterotrophic microflora, and direct counts using acridine orange method (Hobbie et al., 1977) were used to calculate the volume of the cells, according to the formula of Krambeck et al. (1981).

At Arcachon, changes in the qualitative composition of the heterotrophic community were assessed by isolating at random 240 pure bacterial strains from plate counts for each seasonal experiment. Each strain was characterized by one hundred morphological, physiological, biochemical and nutritional tests including cellular morphology, Gram staining, colonial morphology and colour, production of exoenzymes, ability to oxidize or ferment glucose, ability to grow at 4, 37 and 44°C, and ability to reduce nitrate and nitrite. Nutritional tests, ability to grow without NaCl and without growth factors such as vitamines or amino acids were performed on 38 organic compounds as carbon and energy sources. These results are coded in binary form (0 or 1).

The catabolic potentialities of the bacterial communities isolated at each sampling time were described by synthetical average indexes, calculated by averaging the caracteristics of the pure strains composing the communities. These indexes are the average index of exoenzyme presence (EAI, indicating the ability to utilize macromolecules) and the average index of carbonaceous compounds utilization (UAI, indicating the ability to degrade low weight compounds corresponding to the 38 substrates tested as being the sole source of carbon and energy). For each family of substrate, an average index of utilization was calculated : amino acids, carbohydrates, alcohols, fatty acids and organic acids of intermediary metabolism. These indexes were calculated as described by Van Wambeke et al. (1984).

At Roscoff, a similar technique using a reduced number of tests (20) was utilized (api 20B method, see Troussellier and Legendre, 1981). The data obtained by the two methods are not directly comparable, nevertheless they are both suitable to describe the qualitative evolution of the physiological potentialities of the communities studied.

In both localities, the strains were sorted according to their similarities and clustered using the KHI_2 coefficient and variance analysis. On the graphical representation of this sorting (dendrograms), any group or isolate at a level of 70 % (90% for the api 20 B tests) of similarity was considered as a different bacterial ecotype.

On the basis of the binary codification of the responses to the previously described tests, a hierarchical classification was done for each water sample. A partial dendrogram was established for each sample, and the totality of the strains isolated during the experiment were compared in a general dendrogram.

The analysis of the general dendrogram gives a mean to compare different communities and to characterize their degree of individualization. The comparison of the strains isolated at successive sampling times gives an indication of the dynamics of the bacterial communities (gradual when the strains are gathered in the same cluster, or rapid, when the communities isolated at successive sampling times are distinct).

Seasonal patterns in Arcachon and in Roscoff

The experiments were performed on samples collected in Arcachon (September-Summer 1982 and February-Winter 1983) and on samples collected in the Aber of Roscoff (December-Winter 1984). Microcalorimetric responses obtained for Arcachon microcosms displayed the typical summer and winter patterns as described by Tournié & Lasserre (1984). Conversely, at Roscoff, during two consecutive years of experiments (1983-1984), only the typical "summer" and "spring-autumn" patterns were found (i.e. no "winter-type" response was obtained). This stability of the calorimetric patterns might well be related to the more stable temperature regime of the shallow waters in Roscoff (temperature range of 7 to 15°C) than in the Arcachon Bassin or the Gironde estuary (temperature range of 3 to 24°C). The qualitative and quantitative evolution of bacterial populations both in Arcachon and Roscoff were directly related to the microcalorimetric evolution. Equivalent power-time curves obtained in Roscoff (December) and in Arcachon (Septembre) showed the same bacterial dynamics. Conversely, the "winter" calorimetric response obtained in February in Arcachon were related to very different bacterial dynamics.

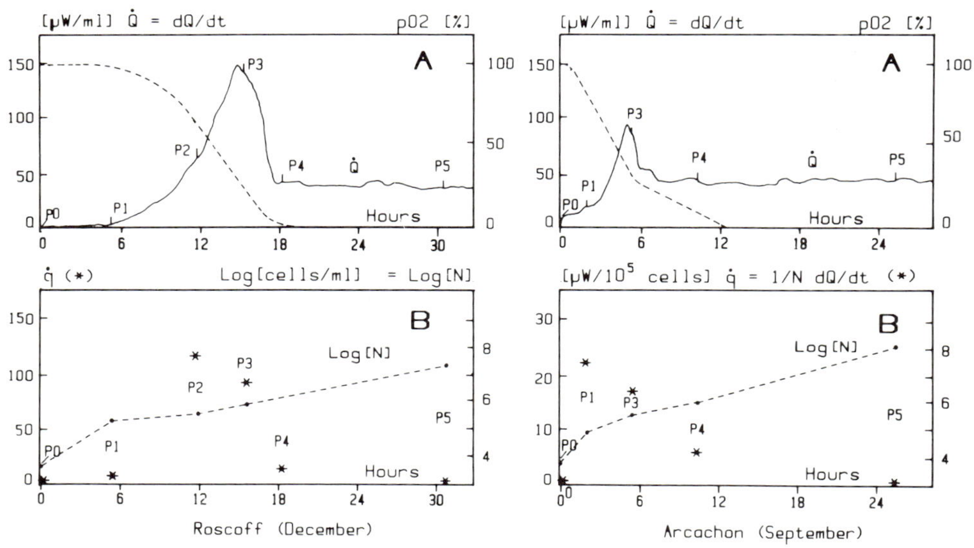

Fig. 4 : Typical "summer" patterns obtained in Roscoff (December 1983) and in Arcachon (September 1982). A : Heat production ($\dot{Q}$) and oxygen tension (pO_2) evolution. B : Specific rate of heat production ($\dot{q}$) and evolution of number of bacteria (N).

In figure 4A, a typical power time curve and the simultaneous oxygen tension variation are shown with the microheterotrophic cell numeration (N : cells/ml). Consequently to the initial organic enrichment (in the microcosm), the heat production increased exponentially and the oxygen concentration decreased proportionally, revealing active aerobic metabolism. In figure 4B, the specific rate of heat dissipation $\dot{q}$ = **1/N dQ/dt** was obtained from the power time curve and the cell numeration data. In both Arcachon and Roscoff, very similar values of $\dot{q}$ were found, the evolution of $\dot{q}$, within the duration of the experiment (30 hours), indicated important variations in energetic capabilities.

One remark can be made refering to the significance of the specific heat dissipation rate (dQ/dt per cell). The nutrient enrichment induced the appearence of an energetically expensive metabolic phase followed by a secondary decrease which indicated a metabolic recovery. This phenomenon was independent of oxygen availability, i.e. in "summer" microcosms where the oxygen tension was severely limited (after enrichment), a similar energetic recovery was observed as in "winter" microcosms (figure 5).

The study of population dynamics showed significant modifications in the physiological potentialities. Before enrichment (time zero) the initial bacterial populations both in Arcachon and Roscoff were characterized by moderate catabolic potentialities (see Table 1), a high diversity and an average cell density of 0.1 and 75 10^5 cell/ml respectively. The cell volume was about 2.6 um^3 in Roscoff.

TABLE 1.

	ARCACHON September			ARCACHON February			ROSCOFF December		
Macromolecules utilization	33	53	42	37	49	33	59	43	82
Organic compounds utlization	26	22	32	36	35	39	45	13	70
Glucose fermentation	25	50	75	65	20	35	0	9	97
NO_3 reduction	75	85	95	70	60	65	39	40	82
	IP	P	FSS	IP	P	FSS	IP	P	FSS

Average percentage of positive responses to 4 types of catabolic and physiological tests. Exoenzymes presence was tested on 7 macromolecules (Arcachon) and on 2 macromolecules (Roscoff). The organic compounds utilization index was calculated on the basis of 34 substrates in Arcachon (amino acids, carbohydrates, alcohols, fatty and organic acids) and on 13 substrates in Roscoff (principally carbohydrates and alcohols). IP : initial bacterial population. P : microcalorimetric peak. FSS : final steady-state.

During the increase of the heat production rate, the succession of populations was very rapid, the strains isolated at different incubation times never clustered at a similarly high level. The values of index of utilization of small molecules (amino acids, carbohydrates, alcohols) showed a net tendency to decrease and reach a minimum indicating physiological specialization, paralleled by the maximum value of the specific heat dissipation rate $\dot{q}$.

During this phase of specialization, the cell volume decreased from 2.6 μm^3 to 1.2 μm^3. In the sample corresponding to the peak, the average volume of the cells is somewhat higher (1.5 μm^3) due to the emergence of a new population of larger cells ($\pm$2 μm^3).

The secondary calorimetric steady state was reached about 3 hours after the peak of heat dissipation. The drastic decrease of the level of $\dot{q}$ was related to the rapid emergence of very different dominant strains, formed by larger cells (2.1 μm^3) and characterized by very broad metabolic potentialities. This homogeneous population (the diversity was almost nil) displayed an adaptation to anaerobic conditions (75 to 100% of the population was able to reduce nitrates or ferment glucose). The cell population increased rapidly and reached values of 9 10^7 cells/ml in Arcachon to 10^8 cells/ml Roscoff. The situation is different in Arcachon, during the winter months (January and February, see figure 5).

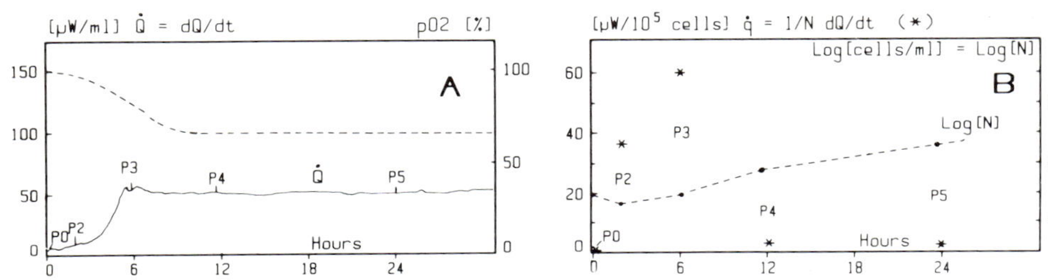

Fig. 5 : Typical "winter" patterns obtained in Arcachon (February 1983). A : Heat production ($\dot{Q}$) and oxygen tension (pO_2) evolution. B : Specific rate of heat production ($\dot{q}$) and evolution of number of bacteria (N).

The initial increase of the heat production $\dot{Q}$ was followed by a secondary steady-state without marked decrease of the heat production (as already described). The final bacterial community dissipated a high level of heat per biomass-unit and the cell population increased slowly from 10^5 cells/ml to 2.5 10^6 cells/ml, 24 hours after nitrogen enrichment. Consequently, the specific rate of heat dissipation $\dot{q}$ reached higher levels than in summer (maximum 60 $\mu W/10^5$ cells, against 20 μW in summer). During the 24-hour experiment, the bacterial diversity was maintained at a high level. This fact indicates that, in spite of an organic input identical to summer enrichment, the qualitative response of the microoheterotroph community was fundamentally different in winter and in summer. Moreover, the evolution of the physiological potentialities of the winter populations was not well marked. It is shown in Table 1 that the average ability to degrade small and large organic molecules remained stable during the experiment. The percentage of bacterial strains able to induce glucose fermentation as well as nitrate reduction showed a tendency to decrease, as a probable result of unrestricted oxygen availability. No characteristic bacterial population appeared at the end of the exponential phase of heat production.

CONCLUSION

Our ability to predict changes in the biogeochemistry of estuaries and lagoons, as well as other shallow water organic-rich environments, is tied to our ability to quantitatively model the dynamics of microbially-mediated energetic changes closely associated with degradation processes and their chemical end products.

Flow microcalorimetry appears well suited for quantifying these energetic changes, in order to reveal microbial phenomena not shown by a more reductionist approach. This is particularly true for complex situations in highly eutrophied coastal systems where bacterial populations play a very important role in the biogeochemical exchanges.

The microcalorimetric data and simultaneous studies of the physiological potentialities on marine bacterial populations, colonising the water-sediment interface, indicated that, at the level of the population dynamics, the power-time curves describes dissipative structures in the form of temporal successions of microorganisms.

The mechanisms underlying power-time curve evolution are due to the succession of bacterial populations with different dynamics and physiological capabilities. "Summer" type microcalorimetric curves are associated with bacterial populations responding to eutrophication by rapid and well marked successions ending in the selection of a community dominated by anaerobes. These anaerobes are characterised by very high catabolic potentialities. The latter are responsible for the observed metabolic recovery shown by a decrease in the heat production, followed by a stable microcalorimetric plateau. During the first part of the microcalorimetric curve (from time zero, PO, to the PTC peak, P3), aerobes gradually increase their specialization (figure 6).

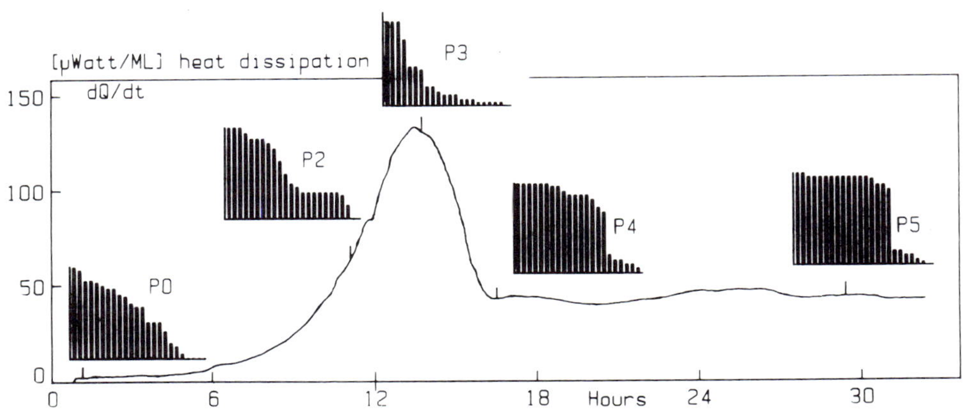

Fig. 6 : Evolution of microbial catabolic potentialities during a "summer" type microcalorimetric response to eutrophication. The five bar charts characterize the specialization of the bacterial population, expressed as the ability to metabolize a variety of organic substrates (carbohydrates, alcohols, amino acids etc.).

When "winter" type microcalorimetric curves are observed, the bacterial successions are less marked. From the begining of the experiment to the secondary steady state, the populations remains very similar and their catabolic potentialities are identical. As a consequence, the specific rate of heat is significantly higher than in summer.

More generally, our studies offer an example of the value of direct microcalorimetry in studying homeostatic properties of ecosystems. Here, heat dissipation with respect to seasons is an adequate parameter of the ability of the system to return, after transitory oscillations, to a new steady state (a parameter of ecological "resilience").

Mixed bacterial populations displayed different energetic regulation of adaptative nature. In effect, the evolution of the specific rate of heat dissipation (1/N dQ/dt) showed a difference in relation to seasonal thermal regimes. A characteristic feature is the increase in the specific heat production rate, shortly after the perturbation (nutrient enrichment) followed by a decrease probably indicating subsequent functional adaptional changes in populations, and marked by a final stationary state. This stationary phase indicates a maximum efficiency with probable adaptation of enzymatic systems to new environmental conditions. According to Schrödinger (1967), the specific rate of heat dissipation is a good index of metabolic efficiency, in living organisms.

Now, one intriging problem is to understand how ecological successions relate to the minimum entropy production based on linear irreversible thermodynamics (Prigogine-Wiame, 1946 ; Prigogine 1982). Recent experimental developments have shown that the thermodynamics of non-equilibrium processes can be used to deduce quantitative relationships and equations of importance to developmental biology (Lamprecht and Zotin, 1978). For Zotin (1985), "the possibility is not to be excluded that thermodynamics might be used to obtain phenomenological equations appropriate to the theory of ecosystems".

ACKNOWLEDGMENTS

This work was supported by the Centre National de la Recherche Scientifique as part of the 'Greco' Interaction Continent-Océan, with additional contributions from the Institut Français de Recherche pour l'Exploitation de la Mer and the Ministère de l'Environnement.

REFERENCES

Beezer, A.E. (ed.), 1980. Biological Microcalorimetry. Academic Press, London and New York, 483 pp.

Belaich, J.P., 1980. Growth and metabolism in bacteria. In: A.E. Beezer (ed.), Biological Microcalorimetry. Academic Press, London, pp. 1-42.

Fenchel, T., 1969. The ecology of marine microbenthos. IV. Ophelia, 6: 1-182.

Fenchel, T., 1970. Studies on the decomposition of organic detritus from turtle grass Thalassia testudinum. Limnol. Oceanogr., 15: 14-20.

Forrest, W.W., 1972. Microcalorimetry. In: J.R. Norris and D.W. Ribbons (eds), Methods in Microbiology, 6B: 285-318.

Gnaiger, E. and Forstner, H. (eds), 1983. Polarographic Oxygen Sensors - Aquatic and Physiological Applications. Springer-Verlag, Berlin, 370 pp.

Gordon, A. and Millero, F.J., 1980. Use of microcalorimetry to study the growth and metabolism of marine bacteria. Thalassia Jugosl., 16: 405-424.

Hobbie, J.E., Daley, R.J. and Jasper, S., 1977. Use of Nuclepore filters for counting bacteria by fluorescence microscopy. Appl. environ. Microbiol., 37: 805-812.

Krambeck, C., Krambeck, K.H.J. and Overbeck, J., 1981. Microcomputer assisted biomass determination of plankton bacteria on scanning electron micrographs. Appl. environ. Microbiol., 42: 142-149.

Lamprecht, I. and Zotin, A.I. (eds), 1978. Thermodynamics of Biological Processes. Walter de Gruyter, Berlin, 428 pp.

Lasserre, P., 1980. Energetic role of meiofauna and epifaunal deposit-feeders in increasing level of microbial activity in estuarine ecosystems, at the water-sediment interface. Actes Colloq. int. CNRS, Paris., 293: 309-318.

Lasserre, P., 1984. The measurement of the enthalpy of metabolism in marine organisms. In: M.J.R. Fasham (ed), Flows of Energy and Materials in Marine Ecosystems, Theory and Practice. Plenum Press, New York, pp. 247-269.

Lasserre P. and Tournié T., 1984. Use of microcalorimetry for the characterization of marine metabolic activity, at the water-sediment interface. J. Exp. Mar. Biol. Ecol., 74: 123-139.

Ljungholm, K., Norén, B., Sköld, R. and Wadsö, I, 1979 a. Use of microcalorimetry for the characterization of microbial activity in soil. Oikos, 33: 15-23.

Ljungholm, K., Norén, B. and Wadsö, I, 1979 b. Microcalorimetric observation of microbial activity in normal and acidified soils. Oikos, 33: 24-30.

Mann, K. H., 1972. Macrophyte production and detritus food chains in coastal waters. Mem. Ist. Ital. Idrobiol., 29 Suppl.: 353-383.

Martens, C.S., 1982. Biogeochemistry of organic-rich coastal lagoon sediments. In: P. Lasserre and H. Postma (eds), Coastal Lagoons. Oceanologica Acta, pp. 161-176.

Mortensen, U., Norén, B. and Wadsö, I., 1973. Microcalorimetry in the study of the activity of microorganisms. Bull. Ecol. Res. Comm. (Stockholm), 17: 189-197.

Nixon, S.W., Pilson, M.E.Q., Oviatt, C.A., Donaghay, P., Sullivan, B., Seitzinger, S., Rudnick, D. and Frithsen, J., 1984. Eutrophication of a coastal marine ecosystem - An experimental study using the MERL microcosms. In: M.J.R. Fasham (ed.), Flows of Energy and Materials in Marine Ecosystems, Theory and Practice. Plenum Press, New York, pp. 105-135.

Oppenheimer, C.H. and Zobell, C.E., 1952. The growth and viability of sixty-three species of marine bacteria as influenced by hydrostatic pressure. J. Mar. Res., 11: 10-18.

Pamatmat, M.M., 1980. The annual mineralization of organic matter in sediments. Present knowledge, persistent technical problems and uncertainities in our measurements. Actes Colloq. int. CNRS, Paris., 293: 309-318.

Pamatmat, M.M., 1984. Measuring the metabolism of the benthic ecosystem. In: M.J.R.

Fasham (ed.), Flows of Energy and Materials in Marine Ecosystems, Theory and Practice. Plenum Press, New York, pp. 223-246.

Prigogine, I., 1982. Physique, Temps et Devenir. Masson, Paris, pp. 275.

Prigogine, I. and Wiame, J.M., 1946. Biologie et thermodynamique des phénomènes irréversibles. Experientia, 2: 451-453.

Schrödinger, E., 1967. What is Life ? The Physical Aspect of the Living Cell. Cambridge University Press, New York.

Spink, C. and Wadsö I., 1976. Calorimetry as an analytical tool in biochemistry and biology. In: D. Glick (ed.), Methods of Biochemical Analyses, 23: 1-159.

Tournié, T. and Lasserre, P., 1984. Microcalorimetric characterization of seasonal metabolic trends in marine microcosms. J. Exp. Mar. Biol. Ecol., 74: 111-121.

Troussellier, M. and Legendre P., 1981. A functional evenness index for microbial ecology. Microb. Ecol., 7:283-296.

Van Wambeke, F., Bianchi, M.A. and Cahet, G., 1984. Short-term bacterial reactivity of nitrogen-enriched sea water of a eutrophic lagoon. Estuarine Coastal Shelf Sci., 19: 291-301.

Wagensberg, J., Castell, C., Torra, V., Rodellar, J. and Vallespinos, F., 1978. Microcalorimetric study of the metabolism of marine bacteria. Detection of rythmical processes. Invest. Pesq., 42: 179-188.

Zotin, A.I., 1985. Thermodynamics and growth of organisms in ecosystems. Can. Bull. Fish. Aquat. Sci., 213: 27-37.

III

UPTAKE OF TRACE ELEMENTS

BY

LIVING ORGANISMS

TRACE METALS - PHYTOPLANKTON INTERACTIONS : AN OVERVIEW*

F.M.M. MOREL

Department of Civil Engineering, R.M. Parsons Laboratory, Massachusetts Institute of Technology, Cambridge, Massachusetts (U.S.A.)

ABSTRACT

Morel, F.M.M., 1986. Trace metals-phytoplankton interactions : an overview. In: P. Lasserre and J.M. Martin (eds), Biogeochemical Processes at the Land-Sea Boundary. Elsevier, Amsterdam.

As shown early in the century, the role of trace metals as nutrients or toxicants to phytoplankton is mediated by the aqueous speciation of the elements. More recently free metal ion activities have been demonstrated to be the parameters determining such physiological effects. It is argued that this result implies no particular role for the free hydrated metal ions but represents simply the thermodynamic tendancy of the elements to react chemically, at pseudo-equilibrium, with cellular functional groups and hence to be bioactive. At present the focus is on disequilibrium processes, particularly the reduction and photoreduction of the metals Fe and Mn at cell surfaces. Such a process may make these essential elements more available to algae. It may also enhance the specificity of the uptake system, other metals which act as coordinative analogs of the essential ones being toxic by interfering with their transport or utilization. These physiological processes correspond to a general ecological view in which aquatic micro-organisms are simultaneously limited in stable systems by several essential elements and their toxic chemical analogs.

HISTORICAL DEVELOPMENTS

The field of trace metal-aquatic microorganism interactions has a long history dating back to the 20's with efforts to culture aquatic organisms, principally phytoplankton, in the laboratory. Based mostly on exhaustive empirical studies, ca. 1960, researchers established the importance of trace element chemistry for the growth of microalgae and pioneered the use of artificial chelating agents in growth medium recipes (Provasoli et al., 1957 ; Droop, 1961 ; Guillard and Ryther, 1962 ; Johnston, 1964). The emphasis was then on the nutritional role of the trace elements and natural and artificial chelators were thought to make iron available to phytoplankton.

---

*This work was supported in part by NOAA grant NA79 AA-D-00077, NSF grant OCE 831 7532, and ONR contract N00014-80-C-0273. A slightly different version of this article was used as part of a final report to NOAA.

Despite some attempts at physiological laboratory studies and a few field experiments, until the mid 70's little progress was made in our mechanistic understanding of the effects of trace elements on microalgae. At that time, fundamental concepts of coordination chemistry were brought to bear quantitatively on the topic ; the focus was put clearly on the issue of chemical speciation in the growth medium and a series of publications demonstrated the central importance of free metal ion activities in determining physiological effects as measured by growth rates (Manahan and Smith, 1973 ; Sunda and Guillard, 1976 ; Anderson and Morel, 1978). Much of the focus was then shifted from the issue of trace element nutrition to that of toxicity ; the role of chelating agents was viewed to be chiefly trace metal--namely copper--detoxification rather than iron nutrition.

This is roughly what the situation was until the late 70's : our understanding of the relevant chemistry rested squarely on coordination equilibrium concepts and physiological studies depended on the use of chelating agents to buffer the chemistry of trace metals. The physiological aspects of the research efforts remained strictly phenomenological ; only integrative biological parameters such as growth rates were being measured and little attention was being paid to the underlying mechanisms.

Based on laboratory results with media of defined chemistry and some early field work, much speculation was being put forth regarding the potential environmental importance of trace metals and chelating agents in aquatic systems. The principal topic of these speculations was the low productivity, nutrient rich, upwelling regions of the oceans which seemed to exhibit the same need for organic matter (i.e. complexing agents) conditioning as was observed in laboratory cultures (Barber, 1973).

THE ISSUE OF FREE ION ACTIVITIES

The early question that dominates the chemistry side of the trace element-aquatic microbiota interaction problem is that of the nature of the relevant chemical species. The demonstrated importance of the free metal ion activities in determining physiological effects has often been interpreted to mean that the hydrated ionic metal species is the active one, toxic or available. This is a profound and widespread misconception partly based on a false analogy with aqueous species such as $HgCl_2$ which is known to control the toxicity of mercury to microorganisms by promoting relatively rapid transmembrane transport. Indeed, the plasma membrane is the principal barrier between the extracellular milieu and the various cellular sites where metals can react and affect cellular physiology. Transport of highly charged cations through the lipid membrane is exceedingly slow, so that the importance of the free metal ion activities cannot in fact be analogous to that of the $HgCl_2$ concentration.

The misconception rests on a confusion between the thermodynamic scale of activities (which measures the tendency of an element to react) and the concentration scale for free, hydrated, ions. Consider for example the not uncommon situation where the free ferric ion activity may be of the order of, say, 10^{-27} M due to hydroxide or chelate formation. Such low activity makes perfect sense as a measure of the reactivity of Fe(III); it hardly represents the actual abundance of the $Fe(H_2O)_6^{3+}$ ion in a one liter solution: one thousandth of a single molecule.

Free metal ion activities thus control biological effects because they determine the reactivity of the metals in systems at equilibrium. Such equilibrium between the external medium and the cell can only be expected to occur with surface sites . The biological effect must then be controlled by the extent of some surface reaction-- either because a surface site is itself the locus of metal action (e.g., inhibition of uptake of some essential element), or because surface binding controls the uptake rate of a metal whose cellular concentration (and hence physiological effect) is thus determined by a balance between accumulation rate and growth. In any case, the critical cellular reaction--say the binding to a transport protein--must be relatively fast compared to the measured biological response so that a state of (pseudo)equilibrium is observed between the organism and the surrounding medium and the physiological response--e.g. the transport rate--is dependent upon the free ion activity. Such a result then implies no particular reaction mechanism ; it does not assign a specific physiological role to the hydrated ionic species; it merely reflects the (pseudo)equilibrium condition in which the extent of all the coordination reactions of the component of interest is determined by its free ionic activity.

From this point of view, the obvious issues that need addressing are those of the achievement of the equilibrium or pseudoequilibrium condition (i.e. the critical chemical kinetics in the medium), of the nature and rate of the slow step in the chain of processes that link the principal metal species to the physiological response (i.e. the actual physiological mechanism of metal action) and, of course, the environmental meaning of it all.

Accordingly, conceptual advances made over the past few years can be classified under three principal headings :

i) Medium chemistry -- our understanding of the aquatic chemical processes interacting with biological responses has expanded to include, in particular, redox and photoredox reactions and the focus has gradually shifted from equilibrium to kinetics.

ii) Physiological responses -- the early phenomenological approach has been replaced by an increasingly mechanistic one with emphasis on the subtle sublethal processes which are most directly relevant for the study of organisms in their natural milieu and on the feedback control processes by which the organisms adjust to the trace element chemistry of the environment.

iii) Ecological implications -- some of the early environmental hypotheses have been tested in field experiments and our understanding of the role of local and global interactions of aquatic microorganisms with trace elements in the economy of marine systems has become both more general and more specific.

MEDIUM CHEMISTRY

Several trace elements are essential for plant growth and many are toxic. Metals such as iron and manganese, which are needed in relatively large quantities and are rather insoluble in aquatic systems owing to the formation of hydrous oxides, are not known to be toxic to aquatic plants. Most reactive metals such as cadmium, lead or mercury for which there are no known algal requirements become toxic at sufficiently high (but still very low) free ion concentrations. Some metals such as zinc and copper can be limiting at very low free ion concentrations and toxic at higher values. From laboratory and field studies, it would appear that iron, manganese and zinc may be nearly limiting in some marine systems, while copper, cadmium and a few others may be approaching toxicity (Huntsman and Sunda, 1980 ; Morel and Morel-Laurens, 1983 ; Morel and Hudson, 1985).

In accordance with the previous discussion, a mechanistic look at the relationship between biological effects and trace metal chemistry in aquatic systems must focus, not on the relative role of various metallic species in solution, but on the equilibrium condition itself. In some instances, we expect some reactions in the medium or at the surface of the organism to be slow, particularly those that involve iron or manganese oxide. (Iron will serve as our chief example throughout this discussion). To examine these possible slow processes and the resulting disequilibrium condition, we must distinguish between laboratory buffered systems in which low free metal ion activities are maintained by strong chelating agents, and the condition more typical of natural systems in which low free metal ion activities reflect mostly low total metal concentrations.

In the first situation, large concentrations of chelated metals are present and precipitates are normally avoided ; in the second, hydrous oxide solids of iron and manganese are usually found and buffering results simply from the relatively low concentrations of organisms in the system. That equilibrium appears to be reached in buffered laboratory systems is rather surprising. The kinetics of chelate dissociation are relatively slow (Table 1) and one would expect it to control the rate of cellular binding over a time scale of hours to days--which is typical of the measured physiological effects. This question merits additional studies ; at this point, since the reactions do appear to be fast, we are left speculating on the possible role of ternary complex formation (chelates-metal-surface ligand), of low surface pH and redox potential, and of photochemistry in accelerating the metal exchange process.

In unbuffered systems, kinetic limitation may come either from slow transport to the cell surface or from slow dissolution of solid species resulting in depletion of the bulk medium. Simple calculations show (Table 1) that, in the open ocean, steady state diffusion of such elements as zinc of iron to the cell surface may match the uptake rate of fast growing algae. This is in accordance with the postulate that, in a stable environment, all potentially limiting elements should effectively be co-limiting primary production and that the average uptake rates should match the diffusion rates (Morel and Hudson, 1985).

TABLE 1

Slow processes in solution (based on Fe uptake by T. weissflogii).

1. Rate of cellular Fe uptake

Minimum for optimum growth : $\rho_{min} = 2x10^{-17}$ mol cell^{-1} hr^{-1}

In typical culture (4 10^6 cell l^{-1} $\cong$ 0.4 mg l^{-1})

$\rho_c = 4x10^{-10}$ mol^{-1} hr^{-1}

2. Steady State Diffusion in Open Surface Ocean

$(Fe)_T \cong 10^{-10}$ M ; cell diameter = 10 μm

$J = 2x10^{-17}$ mol cell^{-1} hr^{-1}

3. Time for soluble Fe depletion in unbuffered culture

$(Fe)_T \leq 10^{-9}$ M $\quad$ $t = (Fe)_T/\rho_c \leq 2.5$ hr

4. Rate of uncatalyzed FeO_x dissolution

$k_d = 10^{-14}$ mol cm^{-2} sec^{-1} (?) ; $(FeO_x)_T = 2x10^{-8}$ M ; area = 3 cm^{2}l^{-1}

$r_d = 10^{-10}$ mol l^{-1} hr^{-1}

5. Rate of exchange from chelator

exchange with transferrine $\quad$ $k_e = 10^{-8}$ sec^{-1} ; (FeY) = 10^{-7} M

$r_e = 2\ 10^{-12}$ mol l^{-1} hr^{-1}

In fact k_e must be at least 100 x greater since r_e matches ρ_c

In unbuffered (i.e. unchelated) culture media, we can calculate that the soluble iron can easily be exhausted in a few hours (Table 1) ; the same may be true of field situations when the cell concentrations are high. Though the precise rates of hydrous iron oxide dissolution are not precisely known, the uncatalyzed process is probably insufficient to keep up with biological uptake. In fact, the limiting rates observed in unbuffered media (Anderson and Morel, 1982) correspond roughly to the value calculated in Table 1 ($4x10^{-10}$ mol l^{-1} hr^{-1} for $(FeO_x)_T = 2x10^{-8}$ M). This result points at the role of chelating agents, especially those that form light absorbing ferric complexes, in making iron available to algae. Certainly organic chelators do enhance the rate of iron oxide dissolution, both through thermal and photochemical reactions (Waite and Morel, 1984a & b). Since the rate of delivery of iron from the chelator to the cell is unexpectedly fast (Table 1), particularly when the complex is photoactive (Anderson and Morel, 1982), we must conclude that the early investigators were right : chelating agents, including natural humic substances, make iron more available to algae. This "kinetic role" of the chelators must be contrasted to their equally important "equilibrium role" in repressing trace metal toxicity.

As we have discussed, non-equilibrium conditions in the medium may be achieved because of some slow transport or reaction rate. Disequilibrium may also result from fast, energy consuming processes that maintain a steady state supply of unstable chemical species such as those formed by photochemical redox reactions (Fig.1). For example, we now know that photo-reduction of metal-chelates enhances the rate of iron uptake by algae (Anderson and Morel, 1982), that organic surface complexes on iron oxide induce photoreductive dissolution of the metal (Waite and Morel, 1984a & b), and that algae themselves promote the photoreduction of iron (Anderson and Morel, 1980). Direct evidence for the role of iron photoreduction in promoting algal iron uptake in natural systems is lacking however and will be difficult to establish owing to the fast kinetics of Fe(II) oxidation. Vice versa, we know little regarding the mechanisms of manganese photoreduction which may well occur as the result of complicated indirect processes since we do not even know the likely chromophores. Nonetheless the net manganese reduction promoted by light and its role in the manganese nutrition of phytoplankton has been demonstrated straightforwardly, thanks to the slow kinetics of Mn(II) reoxidation (Sunda et al., 1982).

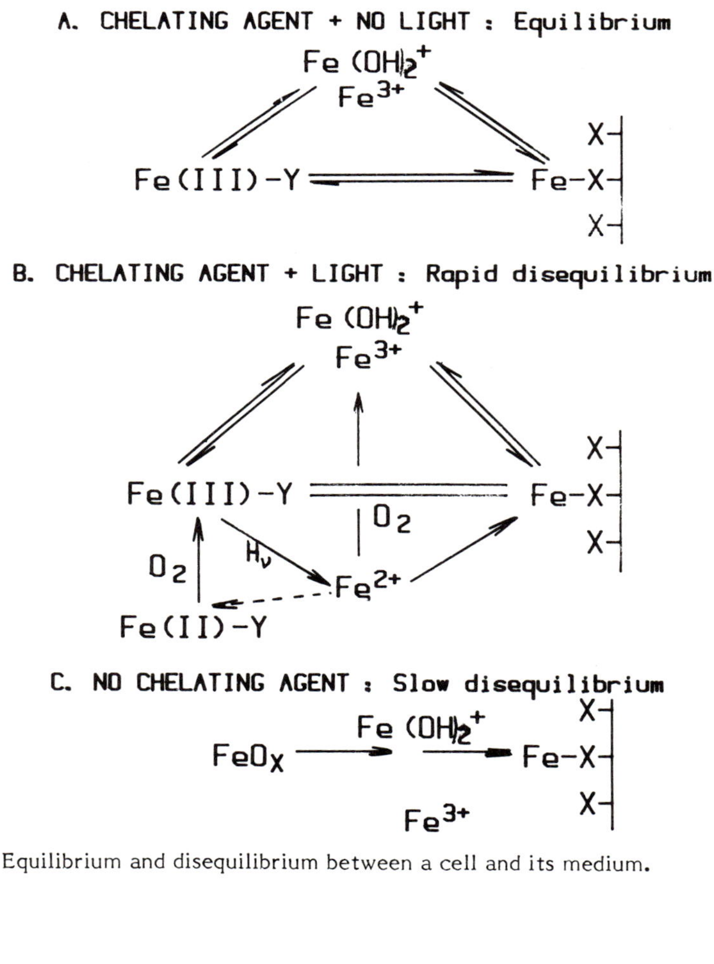

Fig. 1. Equilibrium and disequilibrium between a cell and its medium.

There is truly little doubt at this point that photochemical processes play an important, perhaps a critical role in the uptake of iron and manganese by algae. The situation is a good deal more complex than sketched so far however. The tight association of metal oxide particles and cell surfaces, the existence of large concentrations of surface binding sites not directly involved in metal transport, and the release of reactive redox compounds such as hydrogen peroxide by the algae (Zepp, personal communication), all greatly complicate the interplay among diffusion, coordination, redox, and photochemical processes.

This complex situation is obviously a major focus for future work and it may prove fruitful to develop some working hypotheses to unravel the problem. For example the observed strong metal binding to the algal surface is apparently not directly involved in metal transport and its physiological role is presently unclear (Hudson, R. personal communication). Perhaps such binding may be useful in an environment with regular light/dark periodicity : as illustrated in Fig. 2, photochemically induced reduction and solubilization of iron and manganese (either directly due to reactivity of ferric chromophores or indirectly due to reactivity of MnO_x with H_2O_2) may augment considerably the specificity of the uptake process which would then favor Fe and Mn over the other trace metals that do not possess similar redox properties. In other words one may hypothesize that the (photo)redox properties of the elements are utilized to obtain a specificity of uptake that may be difficult (or costly) to achieve on the basis of coordination alone. Alternatively, the light/dark cycle and the consequent cycle of photosynthesis and respiration may generate a significant diel variation in pH, resulting in alternating periods of iron and manganese deposition and solubilization which may favor uptake.

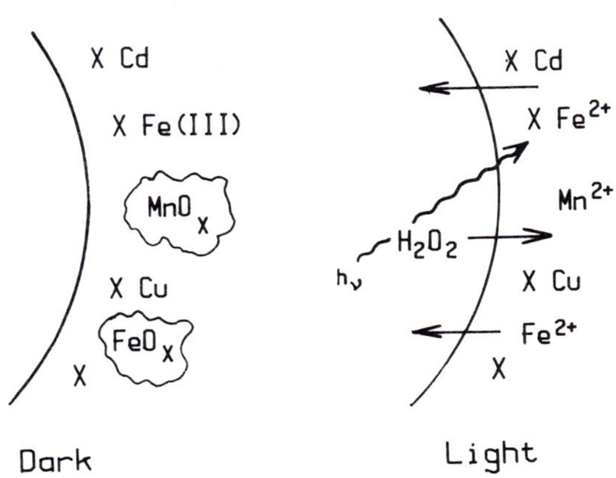

Fig. 2. Hypothesis: The utilization of (photo)redox properties of Fe and Mn at the cell surface augments the specificity of uptake.

PHYSIOLOGICAL MECHANISMS

The question of the physiological mechanisms of metal effects on microorganisms poses some fundamental conceptual and experimental difficulties. The first difficulty is partly a matter of semantics and concerns the notion of "cause", "mechanism" or "explanation". For example, focussing on the issue of metal toxicity, is it a satisfactory explanation to say that cadmium affects growth rate in T. weissflogii by reducing photosynthetic rate ? There is a long chain of intermediary processes linking the primary cause (high Cd^{2+}) to the ultimate phenomenological observation (reduced growth rate) and inhibition of chlorophyll synthesis or of iron nutrition are equally valid explanatory mechanisms. The situation is made more complicated by the tight couplings among the different parts of the cellular machinery and its control by physiological feedback processes. As a result, the integrated growth rate response to the presence of a toxic element typically depends on almost all important environmental variables : temperature, pH, light, major and trace nutrients, etc. Finally, highly reactive trace metal species have multiple effects on living cells. Depending on concentrations (and on physiological and environmental conditions), practically all metal reactive groups, from highly specific metal centers in enzymes, to calcium channels, or unspecific ligands in proteins can be affected by the presence of a foreign metal ion. Corresponding to these different reaction sites, different mechanisms of toxicity will be exhibited by the organism at different metal concentrations. To obviate these difficulties, we wish to focus here, as much as possible, on the initial site of action of a toxic metal, at the lowest concentration (activity) that yields a measurable physiological effect, under environmental conditions that are typical of the organism's natural milieu. For example, it is the primary effect of cadmium in inhibiting iron uptake or assimilation at low cadmium ion activities (Harrison and Morel, 1983) which is of interest to us, not the resulting chlorosis or the potassium loss that can be induced at high (Cd^{2+}). Of course, not all ambiguity is removed in this way. For example, depending on the relative concentrations of the other essential metals, within the normal environmental range, (and possibly on other factors such as light, etc.) cadmium toxicity may also be effected though inhibition of manganese or zinc nutrition--or two or three of these mechanisms concurrently.

The obvious and fundamental concept that results from such elucubrations, one that is now supported, documented and quantified in several experimental data sets, is that the toxic effect of a trace metal is caused by its interference with the uptake, assimilation or utilization of another, essential trace metal. Trace metals are to various degrees coordinative analogs of each other, toxicity occurs when a metal binds competitively to a ligand site involved in the transport, assimilation or utilization of an essential metal whose function it cannot emulate. (In some cases metals can replace others with reasonably good efficiency). As a result, the study of trace metal toxicity in microalgae must commence with a study of trace metal nutrition, a scientifically sound if paradoxical situation.

The algal requirements for three essential trace metals--Fe, Mn and Zn--appear to determine trace metal toxicity in marine phytoplankton. Other essential metals such as copper or cobalt are sufficiently abundant that they more commonly function as toxicants than as limiting elements. Zinc presents the pecularity that it can both be limiting or toxic ; in most phytoplankters, the thresholds of Zn^{2+} limitation and toxicity are fairly close to each other, leaving a relatively narrow range of conditions that result in optimal growth.

A number of case studies have been performed with each of the principal essential metals, iron, manganese and zinc and a variety of toxic metals (e.g. cadmium, copper, nickel) chosen as examples for pragmatic reasons such as high toxicity, simple chemistry, existence of radiotracers, etc. (Rueter and Morel, 1981 ; Sunda and Huntsman, 1983). As an example let us consider the studies of iron and cadmium antagonism in the diatom Thalassiosira weissflogii (Foster and Morel, 1982 ; Harrison and Morel, 1983). In these studies different toxic effects were observed depending on the time scale and on the free iron concentration. Over a time scale of hours, the principal effect of the toxic metal (Cd) is to decrease the uptake of iron whose supply simply becomes insufficient over the long-term if $[Fe^{3+}]$ is low. The situation can be characterized as an equilibrium situation between the medium and the critical transport site inactivated by binding of the toxic metal.

If $[Fe^{3+}]$ (or any affected essential element) is high enough, the uptake rate remains sufficient to support (theoretically) a high growth rate. In this case however, the intracellular buildup of the toxic metal (Cd apparently leaks through the Fe transport system) eventually becomes high enough to inhibit Fe assimilation. The cells then become effectively iron defficient though their cellular iron content is abnormally high. In this case the intracellular concentration of the toxic element is the critical parameter ; it is determined by a balance between growth rate and uptake rate, the latter being fixed by a pseudo-equilibrium with a membrane porter site.

Because the net result of trace metal inhibition can paradoxically be either decreased or increased cellular content of an essential element, such studies require rather detailed physiological and biochemical studies to establish mechanisms. Vice-versa, if one wishes to study trace metal toxicity in the field, gross chemical and biological measurements are clearly insufficient ; one must depend on specific physiological and biochemical markers.

The phenomenon of toxic metal interference with essential metal nutrition is complicated by the physiological control that the cells exercise over the uptake and assimilation of all critical nutrients. Studies of the regulation of iron (Harrison and Morel, in preparation) and manganese (Sunda and Huntsman, 1985) transport in phytoplankton have yielded similar results and the general pattern is consistent with what is known for all essential algal nutrients. Microalgae regulate their trace element uptake and, as expected, they do so by changing their maximum uptake rate depending on their degree of starvation or repleteness for a particular element. In round numbers, the maximum uptake rate of starved cells for Fe or Mn exceeds that of replete cells by a factor of 10 to 100. Together with the variations in cellular iron and manganese quotas, these variations in maximum uptake rates allow high growth rates to be maintained at very low free metal ion concentrations. The resulting half saturation constant for growth (limited either by Fe or Mn) is a factor of 100 to 1,000 lower than the half saturation constant for metal uptake. The feedback control on the uptake mechanism takes effect in a matter of several hours (typically a day or so) and is prone to overshoot after starvation of iron or manganese.

The similarities observed among the regulation and toxicity mechanisms in the various experimental systems provides some confidence that one is studying fundamental physiological processes (almost) universally shared among algae. The sound chemical logic that buttresses the physiological observations bolsters that confidence. In a few years, we should have elucidated the fundamental molecular biochemical processses that underlie these observations. Then we will have the highly specific tools needed to attack efficiently the environmental issues.

ENVIRONMENTAL SIGNIFICANCE

While it is difficult at present to address specific environmental questions (such as whether or not a particular organism or ecosystem is under a particular stress) our improved understanding of microalgae-trace metal interactions provides us with new insights on the role of trace metals in phytoplankton ecology and on the role of phytoplankton in trace metal geochemistry. Though much of the conclusions must be speculative at this point (there have been only few field experiments that could and have been performed) they provide a general framework to view the issue of trace metal pollution in coastal waters.

One central ecological hypothesis is that organisms in their natural milieu should necessarily be at the limit of deficiency of several essential elements and of toxicity of their chemical analogs, simultaneously (Morel and Hudson, 1985). This is a generalization of a concept proposed by Redfield fifty years ago for nitrogen and phosphorus (Redfield, 1934), and is precisely the opposite of Liebig's law of the minimum in which only one element is normally expected to be limiting. One could consider the available evidence that manganese-copper interactions may be limiting productivity in freshly upwelled water (Sunda et al., 1981 ; Sunda and Huntsman, 1983) to provide partial support to the general notion of multiple element limitation and toxicity in the sea. The consequence of such a situation is that any significant environmental modification must result in significant ecological change. For the specific case of trace metals, the situation is of course complicated by the chemistry of the elements (e.g., complexation, adsorption, redox, etc.) and by the normal competitive ("antagonistic") mode of action of toxic metals (e.g., increased Cd may have no effect on Fe nutrition if Fe is increased simultaneously).

The complementary geochemical hypothesis is that binding of metals by phytoplankton (followed by grazing and fecal pellet settling) is the major mechanism for removal of trace metals from the water column. In addition to their geochemical control of the cycle of the trace elements, phytoplankton also exert a chemical control by releasing complexing and (photo)reducing agents and promoting intracellularly a host of chemical transformations, methylation in particular. According to this hypothesis, the trace metal chemistry in coastal waters is under rather tight control by the organisms that it affects. Though dramatic ecological effects may be observed during episodes of extreme conditions of trace metal toxicity, in most situations we should expect the ecosystem to exhibit strong homeostatic behavior. The interdependence of the ecological and geochemical processes affecting phytoplankton ecology and trace metal geochemistry is stylized in Fig. 3.

$$\rho_M = \frac{\{M\}}{K_M + \{M\}} \cdot \rho_{max}$$

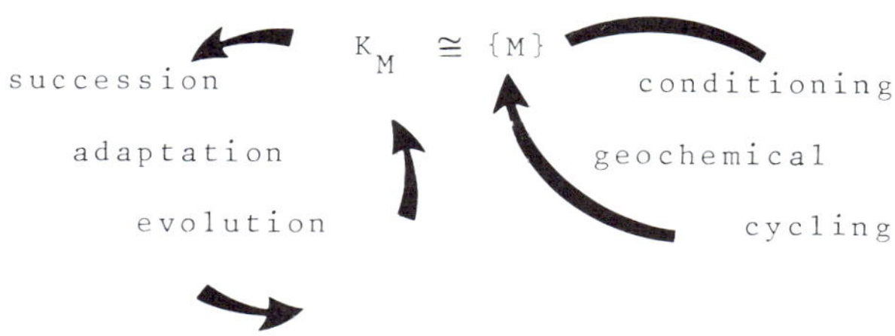

Fig. 3. Interdepedence of aquatic chemistry and biology : The geochemical cycling and the biological effect of a trace element M are both modulated by the uptake rate of M.

Through adaptation, evolution and species succession, the physiology of the organisms is matched to its chemical environment. Conversely, the aquatic chemistry of the critical elements is controlled by geochemical and chemical processes that depend on the physiology of the organisms. The mutual control of the aquatic biology and chemistry can be made more quantitative by considering that the half saturation activity of all biologically active elements should be roughly equal to the element's activity in the medium ; perhaps ten times higher for essential elements ; perhaps ten times lower for the toxic elements that act as their chemical analogs.

REFERENCES

Anderson, D.M. and Morel, F.M.M., 1978. Copper sensitivity of Gonyaulax tamarensis. Limnol. Oceanogr., 23: 283-295.

Anderson, M.A. and Morel, F.M.M., 1980. Uptake of iron II by a diatom in oxic culture medium. Mar. Biol. Lett., 1: 263-268.

Anderson, M.A. and Morel, F.M.M., 1982. The influence of aqueous iron chemistry on the uptake of iron by the coastal diatom Thalassiosira weissflogii. Limnol. Oceanogr., 27: 789-813

Barber, R.T., 1973. Organic ligands and phytoplankton growth in nutrient-rich seawater. In: P.C. Singer (ed.), Trace Metals and Metal - Organic Interactions in Natural Waters. Ann Arbor Science, Michigan, pp 321-338.

Droop, M.R., 1961. Some chemical considerations in the design of synthetic culture media for marine algae. Botanica Marina, 2: 231-246.

Foster, P.L. and Morel, F.M.M., 1982. Reversal of cadmium toxicity in the diatom Thalassiosira weissflogii. Limnol. Oceanogr., 27: 745-752.

Guillard, R.R.L. and Ryther, J.H., 1962. Studies of marine plankton diatoms.1. Cyclotella nana Hustedt and Detonula confervacea (Cleve) Gran. Can. J. Microbiol., 8: 229-238.

Harrison, G.I. and Morel, F.M.M., 1983. Copper sensitivity of Gonyaulax tamarensis. J. Phycol., 19: 495-507.

Huntsman, S.A. and Sunda, W.G., 1980. The role of trace metals in regulating phytoplankton growth with emphasis on iron manganese and copper. In: I. Morris (ed.), The Physiological Ecology of Phytoplankton. Blackwell Scientific, Oxford, pp. 285-328.

Johnston, R., 1964. Seawater, the natural medium of phytoplankton. II. Trace metals and chelation, a general discussion. J. Mar. Biol. Ass. U.K., 44: 87-109.

Manahan, S.E. and Smith, M.J., 1973. Copper micronutrient requirements for algae. Env. Sci. Technol., 7: 829-833.

Morel, F.M.M. and Hudson, R.J.M., 1985. The geobiological cycle of trace element in aquatic systems : Redfield revisited. In: W. Stumm (ed.), Chemical Processes in Lakes. Wiley Interscience,New York, pp. 251-281.

Morel, F.M.M. and Morel-Laurens, N.M.L., 1983. Trace metals and plankton in the ocean : fact and speculations. In: C.W. Wong, E. Boyle, K.W. Bruland, J.D. Burton and E.D. Goldberg (eds), Trace Metals in Sea Water. NATO Conference Series, Plenum Press, New York, pp. 841-869.

Provasoli, L., McLaughlin, J.J.A. and Droop, M.R., 1957. The development of artificial media for marine algae. Arch. Mikrobiol., 25: 392-425.

Redfield, A.C., 1934. On the proportion of organic derivatives in the sea water and their relation to the composition of plankton. In: James Johnston Memorial Volume. Liverpool University Press, Liverpool, pp. 176-192.

Rueter, J.G. Jr., Chisholm S.W. and Morel, F.M.M., 1981. Effects of copper toxicity on silicic acid uptake and growth in Thalassiosira pseudonana. J. Phycol., 17: 270-278.

Sunda, W.G., Barber, R.T. and Huntsman, S.A., 1981. Phytoplankton growth in nutrient rich seawater : Importance of copper-manganese cellular interactions. J. Mar. Res., 39: 567-586.

Sunda, W.G. and Guillard, R.R.L., 1976. The relationship between iron activity and the toxicity of copper to phytoplankton. J. Mar. Res., 34: 511-529.

Sunda, W.G. and Huntsman, S.A., 1983. Effect of competitive interactions between manganese and copper on cellular manganese and growth in estuarine and oceanic species of the diatom Thalassiosira. Limnol. Oceanogr., 28: 924-934.

Sunda, W.G. and Huntsman, S.A., 1985. Regulation of cellular manganese transport rates in the unicellular alga Chlamydomonas. Limnol. Oceanogr., 30: 71-80.

Sunda, W.G., Huntsman, S.A. and Harvey, G., 1983. Photoreduction of Mn oxide in seawater : geochemical and biological implications. Nature, 301: 234-236.

Waite, T.D. and Morel, F.M.M., 1984a. Photoreductive dissolution of colloidal iron oxide : effect of citrate. J. Colloid Interface Sci., 102: 121-137.

Waite, T.D. and Morel, F.M.M., 1984b. Photoreductive dissolution of colloidal iron oxides in natural waters. Environ. Sci. Technol., 18: 860-868.

BIOLOGICAL AVAILABILITY OF TRACE ELEMENTS

D.R. TURNER

Marine Biological Association of the U.K., Citadel Hill, Plymouth PL1 2PB (United Kingdom)

ABSTRACT

Turner, D.R., 1986. Biological availability of trace elements. In: P. Lasserre and J.M. Martin (eds), Biogeochemical Processes at the Land-Sea Boundary. Elsevier, Amsterdam.

The chemical speciation of trace elements in seawater covers a wide range of chemical types from weakly complexed cations though strongly complexed cations to strong-acid anions. Although relatively few elements have been studied, the biologically available chemical species of different elements which have been identified cover the same wide range. The development of chemical techniques capable of measuring these biologically available chemical species is reviewed. While reliable methods have been developed for inorganic and alkylated forms of several metalloid elements, the measurement of ionic species at trace levels requires further development.

INTRODUCTION

Aquatic organisms are continuously undergoing chemical exchanges with their environment, involving uptake and excretion of both nutrient and toxic elements. These processes have important consequences not only for the organisms, but also for the elements concerned, since the vertical profiles of many trace elements in the oceans appear to be biologically controlled (Bruland, 1983). Biological uptake from the dissolved state is of particular interest to chemists since it is strongly influenced by the chemistry of the surrounding aquatic environment. The recognition that the uptake of trace metals by aquatic organisms is a function of the chemical speciation of the elements concerned led to the use of the term biological availability, representing the fraction of the total trace metal available for uptake. Biological availability is thus essentially a chemical concept which largely ignores the complexity of the biological processes involved. Nevertheless it offers a means of relating the chemistry and geochemistry of a trace element, as manifested in its chemical speciation, to its interactions with the biota.

Before considering the details of biological availability it is important to put this parameter in perspective. Long-term uptake is usually described by an enrichment factor equal to the ratio of concentrations of an element in an organism and in seawater. Enrichment factors for a wide range of elements in marine organisms cover 4-5 orders of magnitude, from near unity for Na, Mg, Cl to 10^4 or more for phosphorus and the heavy metals (Trudinger et al., 1979 ; Spaargaren and Ceccaldi, 1984). Despite this range of enrichment factors, the biologically essential elements appear to have been selected simply on the basis of total concentration in seawater. The vertical divisions in Table 1 group the stable elements into three concentration ranges in seawater, and it can be seen that the biologically essential elements correspond to those with concentrations greater than 1nM in seawater, with a small number of exceptions. If we assume that the biologically essential elements correspond to the most available elements with the required range of chemical properties (Williams, 1981) then it is clear that, over large scales of time and space, total concentration in seawater gives an adequate estimate of availability to the biota.

TABLE 1

Inorganic speciation, seawater concentration and biological function of the elements[a].

Speciation[b]	Concentration		
	> 1 uM (nutrient)	1nM - 1uM (micronutrient)	< 1nM (non-essential)
free cation, small amounts of weak complexes	Li,Na,K,Rb, Mg,Ca,Sr	Cs,Ba	
free cations complexes		[Mn(II)],[Fe(II)], Co,Ni,Cu(II),Zn	Y,Cd,La-Lu,[Tl(I)],Pb
complexes, negligible free cation		Al,[Cr(III)], Fe(III),[Cu(I)], [U(IV)], U(VI)	Be,Sc,Ga,Zr,Ag,In,Hf Hg,Tl(III),Bi,Th, [Au(I)], Au(III)
neutral species only		Ti,[Sb(III)]	Sn
neutral species and anions	B, C, Si, P	V,Ge,[As(III)]	Nb,Ta
free anions	N,F,S,Cl,Br	Cr(VI),[Se(IV)] Se(VI),Mo,Sb(V), I, As(V)	W

a : essential elements are underlined ; elements of doubtful status are underlined with dashed lines ; non- equilibrium oxidation states are enclosed in square brackets. Biological function data from Egami (1974) and Trudinger et al. (1979) ; seawater concentrations from Brewer (1975).
b : types of chemical species present at > 1% in seawater (Turner et al., 1981).

The biologically available concentration $[X_b]$ of an element X can be written as :

$$[X_b] = [X_T]\, f_b \qquad (1)$$

where $[X_T]$ is the total concentration and f_b the biological availability or biologically available fraction. It is clear from the preceding discussion that the value of f_b is not significant when one considers the whole periodic table, with a range of $[X_T]$ of at least 12 orders of magnitude, over the evolutionary timescale ; on this scale $[X_T]$ provides an adequate estimate of availability. On small time and space scales, however, the situation is rather different. Here the response of individual organisms or species to changes in the chemistry of a natural water can be important. Marine organisms have only limited defences against increased concentrations of many trace elements, and both non-essential and micronutrient trace elements can become toxic at elevated concentration levels. In such cases it is important to be able to define $[X_b]$ more closely (eqn 1), which in turn means obtaining an understanding of the biological availability f_b in addition to the simpler parameter $[X_T]$.

TRACE ELEMENTS AND CHEMICAL SPECIATION

The correlation between seawater concentration and biological function (Table 1) and the relatively limited range of enrichment factors referred to above has important implications for the chemical nature of biological availability. The maximum range of enrichment factors is about 5 orders of magnitude, of which a considerable part can be attributed to complexation or other sequestration of the heavier elements within the organisms. The variation attributable to chemical speciation outside the organisms can therefore be expected to be considerably less, perhaps 2-3 orders of magnitude. Thus for each element at least one biologically available form must be present to a significant extent in seawater (> 1 % of the total concentration). This in turn implies a wide range of biologically available chemical species as can be seen from the horizontal classification in Table 1. The stable elements in seawater have been divided into six speciation types based on the strength of complexation of an elemental cation since almost all elements in seawater are present in positive oxidation states. At one extreme lie the alkali and alkaline earth cations which form only very weak complexes and are largely present as the free cation, while at the other extreme hypothetical cations such as N^{5+} and Cr^{6+} interact so strongly with water that they are found as strong acid anions in seawater. The halide ions have been included in this latter group for completeness. Although 1% has been chosen as an arbitrary cut-off point, only minor changes would result from cut-offs of 0.1% or 0.01% corresponding to 3 and 4 orders of magnitude range in f_b respectively.

IDENTIFICATION OF BIOLOGICALLY AVAILABLE CHEMICAL SPECIES

The simple definition of biological availability as the fraction of the total trace metal available for uptake by the biota implicitly assumes that all relevant organisms will have similar uptake characteristics. In addition the use of biological availability as a chemical parameter involves the assumption that it can be identified with particular chemical species or groups of species. Perhaps surprisingly many of the studies carried out to date have lent support to these assumptions, though the detailed mechanisms of the uptake processes remain unclear. Some of the inorganic chemical species identified as biologically available are listed in Table 2. It can be seen that much of the work has concentrated on the divalent cations of the first transition series, together with the chemically similar lead and cadmium, where the free cation is generally found to be the biologically available form. Work in this area has been stimulated by the combination of an important group of micronutrient and toxic elements together with readily available techniques for the study of speciation, though these techniques are still far from ideal (see discussion below). Studies in other speciation types have been much more limited, though some results are now available for anion-forming elements, and studies of "acid rain" have stimulated work on the biological availability of aluminium. In practical terms, however, these elements are considerably more difficult to work with than the divalent cations, owing to the lack of chemical techniques for the direct study of speciation in culture media and in natural water samples.

TABLE 2

Bioavailable Inorganic Chemical Species of the Trace Elements.

Speciation	Element	Bioavailable Species	References
free ions + weak complexes	-	-	
free ions + complexes	Cu(II)	free cation	(1)
	Cu(II)	free cation, OH complexes	(2)
	Pb	free cation	(3)
	Ni	free cation	(4)
	Zn	free cation	(5)
	Cd	free cation	(6)
complexes only	Al	$Al(OH)_2^+$	(7)
	U(VI)	OH, CO_3 complexes	(8)
	Hg	Cl complexes	(9)
neutral species	-	-	
neutral species + anions	-	-	
anions	V	$H_2VO_4^-$	(10)
	Cr(VI)	$HCrO_4^-$	(11)

Key to references :

(1) Sunda & Guillard, 1976 ; Anderson & Morel, 1978 ; Sunda & Gillespie, 1979 ; Hodson et al., 1979 ; Petersen, 1982.
(2) Andrew et al., 1977 ; Magnusson et al., 1979 ; Dodge & Theis, 1979.
(3) Schulz-Baldes & Lewin, 1976 ; Merlin & Pozzi, 1977.
(4) Spencer & Nichols, 1983.
(5) Petersen, 1982 ; Anderson et al., 1978 ; Allen et al., 1980.
(6) Sunda et al., 1978.
(7) Helliwell et al., 1983
(8) Pribil & Marvan, 1976
(9) Pentreath, 1976 ; Davies, 1976
(10) Stendahl & Sprague, 1982
(11) van der Putte et al., 1981.

A further group of biologically available chemical species comprises organic compounds of the trace elements, which may be divided into two groups ; organic complexes of cations, and metal and non-metal alkyls (Table 3). These species are mainly neutral molecules, in contrast to the ionic species identified in Table 2, and are thought to be taken up as a consequence of their lipid-solubility (Florence et al., 1983).

It is clear from this brief survey that biologically available chemical species can include free cations, anionic and cationic complexes, and neutral molecules, depending on the element concerned. Such a disparate collection implies a range of different uptake mechanisms ; probably a minimum of three for positive, neutral, and negative charge types. Furthermore, the estimation of biological availability in terms of chemistry becomes more difficult, since not only will different techniques be required for different elements, but some elements are already included in more than one species type in Tables 2 and 3 (e.g. Cd, Hg, Cu).

TABLE 3

Bioavailable Organic Chemical Species of the Trace Elements.

Type of Species	Element	Bioavailable Chemical Species	Reference
metal-organic complexes	Cu(II)	citrate, oxine, ethylenediamine, diethyldithiocarbamate complexes	(1)
	Cd	diethyldithiocarbamate complex	(2)
	Fe(III)	siderophore (hydroxamate) complex	(3)
metal alkyls	Pb	tetramethyllead, tetraethyllead	(4)
	Hg	methylmercury (?hydroxide complex)	(5)
	Sn	various	(6)

Key to References :

(1) Guy & Ross Kean, 1980 ; Florence et al., 1983.
(2) Poldoski, 1979.
(3) Murphy et al., 1976
(4) Marchetti, 1978 ; Wong et al., 1981
(5) Roderer, 1983
(6) Zuckerman et al., 1978.

CHEMICAL ESTIMATION OF BIOLOGICAL AVAILABILITY

Knowledge of the biologically available chemical species of a particular element will be valuable only if those species can be reliably measured at natural concentration levels. Since it is only very recently that total levels of many elements have been reliably measured in seawater, it is clear that measurement of a chemically defined fraction of that total presents a major analytical challenge. In addition to achieving the necessary detection limit and guarding against contamination it is necessary to match the characteristics of the speciation-sensitive technique with the species to be measured. A wide range of speciation-sensitive techniques is now available including electroanalysis, chromatography, dialysis and ion-exchange. Further development will, however, be required since many techniques are subject to interferences and artifacts when used at low concentration levels in natural waters, and techniques are not yet available for all of the biologically available species so far identified. A detailed discussion of this problem can be found elsewhere (Turner, 1984), so only a brief summary will be given here.

A key problem common to almost all speciation-sensitive measurements is that of kinetics. Speciation-sensitive measurements can be divided into equilibrium, kinetic and destructive techniques. Equilibrium techniques cause no disturbance to chemical equilibria in the sample and result in a true equilibrium measurement ; ion-selective electrode potentiometry is the most familiar example. Kinetic techniques in contrast involve depletion of the species being measured in the region of the sensing surface, thus disturbing chemical equilibria in this region, and may thus produce a larger response than the corresponding equilibrium technique ; no significant changes in the bulk solution composition occur during these measurements. All polarographic and voltammetric techniques fall into this class. Destructive techniques involve the removal of a significant quantity of metal from the bulk of the sample ; ion-exchange, chromatographic and chemical reaction techniques are included here. In the last two cases the response obtained depends not only on the selectivity of the sensor or procedure used but also on the kinetics of the equilibria linking the active species sensed with other chemical species. In the limit of very slow kinetics all techniques will give an equilibrium response since no reactions will occur during the measurement, while at the other extreme of very fast kinetics both kinetic and destructive techniques will give a response corresponding to the total metal in the sample. Between these two extremes the response will depend on

TABLE 4

Measurement Techniques for Bioavailable Chemical Species.

Species	Technique	Nature	Remarks
free cation	potentiometry	equilibrium	limited to total conc $< 10^{-6}$M
	polarography	kinetic	) reactions are
	voltammetry	kinetic	) rapid, so these do
	dialysis	destructive	) not truly
	ion-exchange	destructive	) measure free ions
inorganic complexes	speciation modelling	equilibrium	requires good data for solution composition and stability constants
organic complexes	solvent extraction	destructive	) reactions slower here,
	chromatography	destructive	) so fewer kinetic problems
metal alkyls	hydride generation	destructive	) reactions very slow,
	solvent extraction	destructive	) so no kinetic
	chromatography	destructive	) problems

the relative timescales of the reaction kinetics and the measurement process, and may often be difficult to interpret directly in terms of speciation.

The evidence obtained on biological uptake so far shows that it is an equilibrium process, i.e. the rate of uptake is too slow to disturb any trace metal equilibria even where reaction rates are rapid. The biologically available chemical species identified to date (Tables 2 and 3) can be summarized in four groups (Table 4), with different possibilities for measurement in each group. Perhaps the most successful techniques are those based on hydride generation (Andreae, 1983) which are applicable to a range of hydride-forming metals and metalloids. Here the interconversion reactions between inorganic and alkylated forms of the elements are so slow that they can be ignored ; and a destructive chemical technique (hydride generation) provides an effective method of speciation analysis. As we move up Table 4, reaction rates become faster, and kinetic responses become potentially more significant. This is particularly true in the case of polarographic and voltammetric techniques where the kinetic response due to labile complexes can constitue the major part of the measured response.

CONCLUSION

Our knowledge of biological availability has been developing rapidly over the past few years, and in the process the subject has become considerably more complex. The simple conclusion "biological availability = free metal cation" which arose from much of the earlier work can be seen to be the result of a fortuitous selection of elements for study, and the equally fortuitous avoidance of lipid-soluble organic complexes. It is now clear that we must consider a much wider range of biologically available chemical species in order to include the full range of trace element types shown in Table 1. It is to be hoped that, in common with other areas of scientific work, a naively simple childhood and an adolescence of ever-increasing complexity will be followed by a maturity whose elegant simplicity derives from a clear understanding of the processes involved. Some indications of the nature of that future simplicity can perhaps be seen already. Inspection of Table 2 shows that the charged chemical species which have been identified as biologically available are either free metal cations or singly charged complex ions. In the case of neutral species, lipid solubility has already been mentioned in connection with the uptake of organic complexes, but there are indications that neutral inorganic complexes may be taken up in the same way. Gutknecht and Walter (1981) report that the neutral complex $HgCl_2$ diffuses rapidly through bilayer lipid membranes, and Simkiss (1983) has shown that trace metals forming neutral inorganic complexes (Cu, Zn, Hg, Cd) can be extracted from seawater into a lipid phase, while the more weakly complexed Na, Sr, and Mn cannot. The significance of these results for the biological uptake of trace metals is not yet clear, since free cation uptake of Cu, Zn and Cd is clearly dominant in most of the cases studied (Table 2). In contrast mercury, which is present in seawater only as complexes, must be taken up as either a neutral or charged complex.

Taking the elements as a whole, we may expect the charge and size of ions and molecules to emerge as the major factors controlling biological availability. Chemical methods for the estimation of biological availability will have to be selected or developed for each element from a knowledge of its chemical and biological interactions.

REFERENCES

Allen, H.E., Hall, R.H. and Brisbin, T.D., 1980. Metal speciation - effects on aquatic toxicity, Environ. Sci. Technol., 14: 441-443.

Anderson, D.M. and Morel, F.M.M., 1978. Copper sensitivity of Gonyaulax tamarensis. Limnol. Oceanogr., 23: 283-295.

Anderson, M.A., Morel, F.M.M. and Guillard, R.R.L., 1978. Growth limitation of a coastal diatom by low zinc activity. Nature, 276: 701.

Andrew, R.W., Biesinger, K.E. and Glass, G.E., 1977. Effects of inorganic complexing on the toxicity of copper to Daphnia magna. Water. Res., 11: 309-315.

Andreae, M.O., 1983. The determination of the chemical species of some of the "hydride elements" (arsenic, antimony, tin and germanium) in seawater : methodology and results. In: C.S. Wong, E.D. Boyle, K.W. Bruland, E.D. Goldberg and J.D. Burton (eds). Trace Metals in Seawater. Plenum Press, New York, pp. 1-19.

Brewer, P.G., 1975. Minor elements in seawater. In: J.P. Riley and G. Skirrow (eds), Chemical Oceanography. Academic Press, London, 2nd ed., vol. 1, pp. 415-196.

Bruland, K.W., 1983. Trace elements in seawater. In: J.P. Riley and R. Chester (eds), Chemical Oceanography. and Chester, R. (eds). Academic Press, London, Vol. 8, pp. 157-220.

Davies, A.G., 1976. An assessment of the basis of mercury tolerance in Dunaliella tertiolecta. J. Mar. Biol. Ass. U.K., 56: 39-57.

Dodge, E.E. and Theis,T.L., 1979. Effect of chemical speciation on the uptake of copper by Chironomus tentans. Environ. Sci. Technol., 13: 1287-1288.

Egami, F., 1974. Minor elements and evolution. J. Mol. Evol., 4: 113-120.

Florence, T.M., Lumsden, B.G. and Fardy, J.J., 1983. Evaluation of some physicochemical techniques for the determination of the fraction of dissolved copper toxic to the marine diatom Nitzschia closterium. Anal. Chim. Acta, 151: 281-295.

Gutknecht, J. and Walter, A., 1981. Inorganic mercury (Hg^{2+}) transport through lipid bilayer membranes (Abstract). Biophys. J., 33: 110 a.

Guy, R.D. and Ross Kean, A., 1980. Algae as a chemical speciation monitor - I. A comparison of algal growth and computer calculated chemical speciation. Water Res., 14: 891-899.

Helliwell, S., Batley, G.E., Florence, T.M. and Lumsden, B.G., 1983. Speciation and toxicity of aluminium in a model freshwater. Environ. Technol. Letts., 4: 141-144.

Hodson, P.V., Borgman, U. and Shear, H., 1979. Toxicity of copper to the aquatic biota. In: J.O. Nriagu (ed.), Copper in the Environment. Wiley, New York, Part II, pp. 307-372.

Magnusson, V.R., Harriss, D.K., Sun, M.S., Taylor, D.K. and Glass, G.E., 1979. Relationships of activities of metal-ligand species to aquatic toxicity. In: E.A. Jenne (ed.), Chemical Modelling in Aqueous Systems. ACS Symposium Series 93, Am. Chem. Soc., Washington DC, pp. 635-656.

Marchetti, R., 1978. Acute toxicity of alkyl leads to some marine organisms, Mar. Poll. Bull., 9: 206-207.

Merlini, M. and Pozzi, G., 1977. Lead and freshwater fishes, Part 2 - ionic lead accumulation. Environ. Poll., 13: 119-207.

Murphy, T.P., Lean, D.R. and Nalewajko, C., 1976. Blue-green algae : their excretion of iron-selective chelators allows them to dominate other algae. Science, 192: 900-902.

Pentreath, R.J., 1976. The accumulation of mercury by the thornback ray Raja clavata L. J. Exp. Mar. Biol. Ecol., 25: 131-140.

Petersen, R., 1982. Influence of copper and zinc on the growth of a freshwater alga, Scenedesmus quadricauda : the significance of chemical speciation. Environ. Sci. Technol., 16: 443-447.

Poldoski, J.E., 1979. Cadmium bioaccumulation assays. Their relationship to various ionic equilibria in Lake Superior water. Environ. Sci. Technol., 13: 701-706.

Pribil, S. and Marvan, P., 1976. Accumulation of uranium by the chlorococcal alga Scenedesmus quadricauda. Arch. Hydrobiol. Suppl., 49: 214-255.

Putte, I. van der, Lubbers, J. and Kolar, Z., 1981. Effect of pH on uptake, tissue distribution and retention of hexavalent chromium in rainbow trout (Salmo Gairdneri). Aqua. Toxicol., 1: 3-18.

Roderer, G., 1983. Differential toxic effects of mercuric chloride and methylmercuric chloride on the freshwater alga Poterioochromonas malhamensis. Aqu. Toxicol., 3: 23-34.

Schulz-Baldes, M. and Lewin, R.A., 1976. Lead uptake in two marine phytoplankton organisms. Biol. Bull., 150: 118-127.

Simkiss, K., 1983. Lipid solubility of heavy metals in saline solutions. J. Mar. Biol. Ass. U.K., 63: 1-7.

Spaargaren, D.H. and Ceccaldi, H.J., 1984. Some relations between the elementary chemical composition of marine organisms and that of seawater. Oceanologica Acta, 7: 63-76.

Spencer, D.F. and Nichols, H.H., 1983. Free nickel ion inhibits growth of two species of green algae. Environ. Poll. Ser. A, 31: 97-104.

Stendahl, D.H. and Sprague, J.B., 1982. Effects of water hardness and pH on vanadium lethality to Rainbow Trout. Water Res., 16: 1479-1488.

Sunda, W.G., Engel, D.W. and Thuotte, R.M., 1978. Effects of chemical speciation on toxicity to grass shrimp, Paleometes pugio : importance of free cadmium ion. Environ. Sci. Technol., 12: 409-413.

Sunda, W.G. and Gillespie, P.A., 1978. The response of a marine bacterium to cupric ion and its use to estimate cupric ion activity in seawater. J. Mar. Res., 37: 761-777.

Sunda, W.G. and Guillard, R.R.L., 1976. The relationship between cupric ion activity and the toxicity of copper to phytoplankton. J. Mar. Res., 34: 511-529.

Trudinger, P.A., Swaine, D.J. and Skyring, G.W., 1979. Biogeochemical cycling of elements - general considerations. In: Biogeochemical Cycling of Mineral-forming Elements. Elsevier, Amsterdam, pp. 1-27.

Turner, D.R., 1984. Relationships between biological availability and chemical measurements. In: H. Sigel (ed.), Metals Ions in Biological Systems. Marcel Dekker, New York, Vol. 18, pp. 137-164.

Turner, D.R., Whitfield, M. and Dickson, A.G., 1981. The equilibrium speciation of dissolved components in freshwater and seawater at 25°C and 1 atm pressure. Geochim. Cosmochim. Acta, 45: 855-881.

Williams, R.J.P., 1981. Natural selection of the elements. Proc. Roy. Soc. Lond. B., 213: 361-397.

Wong, P.T.S., Chau, Y.K., Kramar, O. and Bengert, G.A., 1981. Accumulation and depuration of tetramethyllead by rainbow trout. Water. Res., 15: 621-625.

Zuckerman, J.J., Reisdorf, R.P., Ellis, H.V. and Wilkinson, R.R., 1978. Organotins in biology and the environment. In: F.E. Brinckman and J.M. Bellama (eds), Organometals and Organometalloids. ACS Symposium Series, 82, Am. Chem. Soc., Washington DC, pp. 388-422..

PREDICTIONS FOR THE MOBILITY OF ELEMENTS IN THE ESTUARINE ENVIRONMENT

B.R. FOLSOM and J.M. WOOD

Gray Freshwater Biological Institute, University of Minnesota, P.O. BOX 100, Navarre, MN 55392 (U.S.A.)

ABSTRACT

Folsom, B.R. and Wood, J.M., 1986. Predictions for the mobility of elements in the estuarine environment. In: P. Lasserre and J.M. Martin (eds), Biogeochemical Processes at the Land-Sea Boundary. Elsevier, Amsterdam.

All chemical and physical processes, including biologically mediated ones, are governed by thermodynamic and kinetic principles. The composition and distribution of elements at the surface of the earth are far from those defined by equilibrium conditions. Stored potential and chemical energy is released as the various reactions proceed toward equilibrium states. Biological catalysis allows organisms to extract some of the energy from these spontaneous but slow processes. Living organisms themselves are thermodynamically unstable and require significant amounts of energy to allow for the production, compartmentalization, uptake and secretion on the many chemicals needed, along with energy demands for mobility, growth and reproduction. Due to the complexity of most natural systems, thermodynamic and kinetic models are difficult to use in describing the cause and effect of stresses on a specific system because of insufficient data concerning all relevant chemical species and reactions under the natural conditions of pH, pE, temperature and ionic strength.

In part, some simplifications and general observations can be made which allow for less complete, but manageable modeling of natural systems under stress. Trends have been observed and documented for a number of changes in the physical environment. First, changes in pH can dramatically affect the solubility of many metal complexes. Acidification leads to the dissolution of Al, Mn , Fe and Zn from sediments and particulates in significant amounts. Changes in reduction potential genreally correlates to the dissolved oxygen content of the water which is mediated biologically through photosynthesis and respiration. Fe and Mn oxides/hydroxides become soluble under anoxic conditions with most metal ions being insoluble under oxic conditions. Another physical transition occurs in the estuarine environment. Water salinity increases as fresh water mixes with salt water. Transition metals bound to refractory organic materials, sediments and clays can be displaced by the fast exchange alkali and alkaline earth metals or can be recomplexed by inorganic ligands. Therefore, when predicting the effects resulting from changes in the physical environment on elemental mobility, general trends which follow established princip-les can be used as a guide.

Additional effects can be predicted dependent upon the actions of the biological organisms which reside in the particular system. The ability of various organisms to compete is dependent upon the environment including the availability, chemical speciation and distribution of essential elements along with the physical characteristics of light, temperature, pH, pE and ionic strength. In turn, living organisms will modify their environment as a result of cellular metabolism. Photosynthetis leads to the consumption of CO_2 and nitrogen compounds with the production of oxygen, reduced organic compounds and an increase in pH. Respiration results in the consumption of oxygen, nitrate, sulfate and/or oxygen, the degradation of organic molecules to CO_2 with related changes in pE and pH. Because of the requirements for O, C, N, S and P, biological activity, metal mobility and the bulk elemental cycles are intimately linked.

INTRODUCTION

It is understood that chemical species in the biosphere may move from its source through the physical environment and into the biota where it may cause either a beneficial or a toxic effect at the cellular level. The pathways taken by each element involve many chemical species with their distribution being determined by thermodynamic and kinetic considerations. Of primary importance to an understanding of elemental cycling is a knowledge of ion complexation in water, since it is the ubiquitous solvent for life and the major solvent for the movement and reaction of chemicals at the surface of the earth. Trace metals in aqueous solution are seldom free ions, but exist as ionic complexes utilizing a variety of organic, inorganic and hydrated ligands which affect mobility, reactivity and solubility. Compounds which are stable thermodynamically, can often be isolated and quantified. Because of the inadequacies in current analytical methods to separate and quantify individual chemical species, elemental cycling has been limited to monitoring either total elemental abundance or the abundance of stable chemical structures.

From the standpoint of metabolism, elements can be considered to be : 1) non-critical, in that they rarely elicit harmful effects either because they are essential to biological systems or they are rarely observed in harmful concentrations ; 2) toxic and relatively accessible or ; 3) toxic but very insoluble or very rare (Table 1). Either non-critical or essential elements can elicit harmful effects if the availability is too great (Fi. 1).

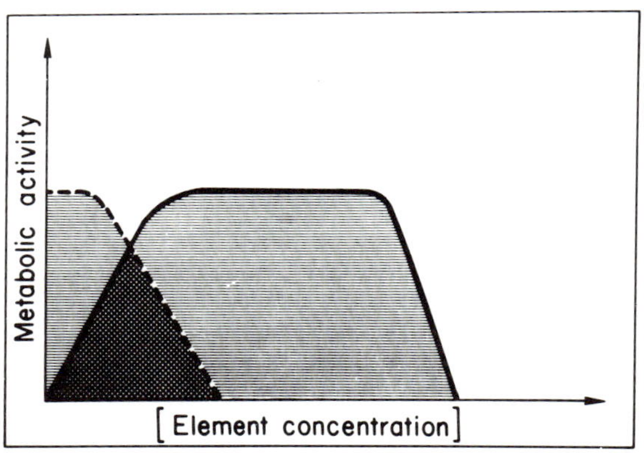

Fig. 1. Metabolic activity vs. concentration for essential and non-essential elements reprinted from Wood and Wang, 1983.

Over the past 200 years since the industrial revolution, we have witnessed the utilization of vast quantitites of energy to mobilize elements at the surface of the earth. Due to these activities, the abundance of many elements in our waterways has increased, leading to the expression of toxic effects and/or the stimulation of biological activity, severely disturbing environmental balances. Intrinsic to an understanding of the ramifications of element mobilization on biological systems is a need to understand the cycling of elements within aqueous systems.

TABLE 1.

Classification of elements according to their toxicity.

Designation[(a)]	Elements
Non critical elements	Na, K, Mg, Ca, H, O, N, C, P, Fe, S, Cl, Br[(b)], F[(b)], Li Rb, Sr, Ba, Al, Si
Very toxic and relatively accessible	Be, Co, Ni, Cu, Zn, Sn, As, Se, Te, Pd, Ag, Cd, Pt, Au, Hg, Tl, Pb, Sb, Bi
Toxic but very insoluble or very rare	Ti, Hf, Zr, W, Nb, Ta, Re, Ga, La, Os, Rh, Ir, Ru

(a) Many elements are omitted because they cannot be designated in this way.
(b) Some may argue with this designation, but we do add F^- to drinking water.
Reprinted from Wood, 1974.

Of the elements in the periodic table, 30 have been shown to be essential for microbial growth, although not all of these 30 elements are necessary for the growth and division of every microbial species (Fig. 2). There is an apparent correlation between the abundance of elements in the earth's crust and the nutritional requirements of microbial cells (Wood, 1984). The rates at which elements are cycled in an aqueous environment depends upon the availability of the bulk elements C, H, N, O, P and S (Salomons and Baccini, 1984).

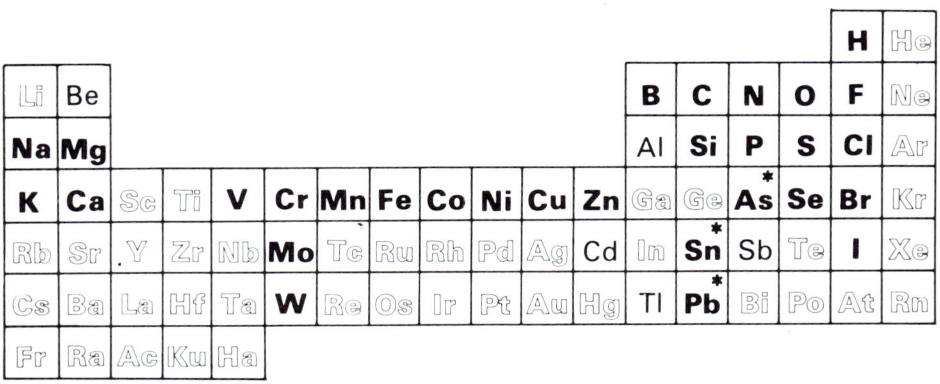

Fig. 2. Periodic table indicating relative abundance and biological importance of elements. Highlighted elements are the 30 most abundant crustal elements with concentrations ranging from 46 % for oxygen to 0.1 µg/g for selenium. Bold faced elements have known biological functions with * indicating limited data and controversy. Data taken from Wood and Wang, 1983.

Not only do microorganisms have a profound influence on the concentration of elements in the atmospheric, aquatic and terrestrial environments (Lovelock, 1979), but the concentration and chemical speciation of these elements predicates which organisms can compete and survive within a given environment. Therefore, the cycling and mobility of trace elements will be linked to the cycling of bulk elements and correlations to the various bulk element cycles can be postulated.

THERMODYNAMIC AND KINETIC AFFECTS

Elements can often be found in any number of chemical and physical states. One common compound formed from 2 hydrogens and 1 oxygen is water. As commonly known, water can exist as a solid, liquid or gas, depending upon the temperature and pressure. Also well known is the fact that water ionizes to OH^- and H_3O^+ in solution. Other combinations of hydrogen and oxygen result in H_2, O_2, H_2O_2, H, O_3,O_2^- and so on. Some of these chemical species are stable whereas others are highly reactive. Such a simple and common combination of two elements leads to a complex and heterogeneous mixture of chemical compounds. When considering that aquatic systems are composed of literally hundreds of stable and reactive chemicals, it is no wonder that researchers have difficulty in quantification and identification.

The abundance and reactivity of any compound is governed by thermodynamic and kinetic principles. Therefore, analytical data is used in conjunction with thermodynamic and kinetic parameters to model complex systems. Often, the most toxic elemental forms are the most reactive and therefore represent the lowest concentrated chemicals species in water. This leads to additional problems in isolation, identification and quantification when assessing potential pollution problems.

Thermodynamic principles govern reactions at equilibrium and indicate the directions of spontaneous reaction. For example, at pH 7.0, ammonia is spontaneously oxidized to nitrate under aerobic conditions by two organisms Nitrosomonas and Nitrobacter.

$$NH_4^+ + 3/2O_2 \text{<-->} + 2H^+ + H_2O \qquad \Delta G'(pH\ 7) = -272\ kJ/mol \qquad (1)$$

$$No_2^- + 1/2O_2 \text{<-->} NO_3^- \qquad \Delta G'(pH\ 7) = -76\ kJ/mol(2) \qquad (2)$$

By contrast, under anaerobic conditions, the reverse reaction becomes spontaneous and organisms such as Micrococcus denitrificans and Thiobacillus denitrificans can reduce nitrate as an energy source.

$$2\ NO_3^- + 5H_2 + 2H^+ \text{<-->} N_2 + 6H_2O \qquad \Delta G'(pH\ 7) = -561\ kJ/mol\ (nitrate) \qquad (3)$$

$$6\ NO_3^- + 5S + 2H_2O \text{<-->} 5SO_4^{2-} + 3N_2 + 4H^+ \qquad \Delta G'(pH\ 7) = -455\ kJ/mol\ (nitrate) \qquad (4)$$

These reactions are spontaneous under standard reaction conditions of 1 M concentrations for all reactants and products (except (H^+) which is at 10^{-7} M). Concentrations of 1 M are far from realistic in most natural waters, therefore, to determine if a reaction will be spontaneous under real conditions, the free energy can be calculated :

$$\Delta G = \Delta G' + RT \ln [prod]/[reac] \qquad (5)$$

Under natural conditions, these organisms derive little energy from nitrate reduction so the total biomass increase is small and slow in comparison to the turnover of reactants. Equilibrium models from thermodynamic calculations can often provide a useful first approximation for a real system. These models are limited by the accuracy and availability of thermodynamic data including temperature, pressure and activity dependencies, knowledge of all pertinent chemical species and related equilibria.

One major assumption, which is often false, is that the spontaneous reactions reach equilibrium quickly. Many reactions in aquatic systems are irreversible and slow with reaction halftimes of minutes or longer. Included in this group of reactions are ; metal-ion oxidation, oxidation of sulfides by organics, various metal ion polymerizations, aging of hydroxide and oxide precipitates, precipitation of metal-ion silicates and carbonates and non-biological hydrolysis of polymers (Stumm and Morgan, 1981). Not only do thermodynamic constants shift with changes in the physical properties of the solution, but kinetic constants vary also. For example, the uncatalyzed oxidation of iron(II) below pH 4 has a reaction halftime of over 3 years, whereas at about pH 7.5 the reaction halftime is in minutes (Stumm and Morgan, 1981). For these reasons, accurate modeling of natural systems becomes more difficult. One approximation can be made for some systems. Usually, one reaction step is slow or rate limiting and therefore all other related reactions can be assumed to be at equilibrium. This leads to a mixed model where the slow reaction is described kinetically with all other reactions being defined by thermodynamic parameters and mass balance constraints.

COMPLEX SOLUBILITY AND MOBILITY

Metal/ligand identity determines whether various chemical species are soluble or insoluble in water. Iron commonly exists in two oxidation states, Fe (II) and Fe(III). In aqueous solution, Fe(II) can be ligated by sulfide ion to form an insoluble iron sulfide complex. Fe(III) can combine with three hydroxide ions to form the insoluble complex, $Fe(OH)_3$, which in turn can coordinate to an additional hydroxyl and form a soluble $Fe(OH)_4^-$ complex. Positively charged metal ions form interactions with free electron pairs from ligands. Multidendate ligands such as those found in refractory organic materials and particulates generally form more stable metal ligand complexes. Metal ions generally have an even number of coordination sites, frequently 4 or 6. The hard soft metal ligand theory states that combinations of metals and ligands are more stable if they are alike, i.e. hard-hard, soft-soft and intermediate-intermediate (Table 2).

TABLE 2

Classification of hard and solft acids and bases(a)

Hard acceptor	Intermediate	Soft acceptor
H^+, Na^+, K^+, Be^{2+}, Mg^{2+} Ca^{2+}, Mn^{2+}, Al^{3+}, Cr^{3+} Co^{3+}, Fe^{2+}, As^{3+}	Fe^{2+}, Co^{2+}, Ni^{2+} Cu^{2+}, Zn^{2+}, Pb^{2+}	Cu^+, Ag^+, Au^+, Tl^+, Hg^+, CH_3Hg^+
Hard donor	**Intermediate**	**Soft donor**
H_2O, OH^-, F^-, Cl^- PO^{3-}, $SO_4{}^{2-}$, $CO_3{}^{2-}$, O^{2-}	Br^-, $NO_2{}^-$, $SO_3{}^{2-}$	SH^-, S^{2-}, RS^-, CN^-, SCN^-, CO, R_2SRSH

(a) Source : Pearson, 1968 ; Forstner and Wittman, 1979.

Metal ligand complexes can be divided into three groups based upon the stability of the complexes and the identity of the preferred ligands (Buffel, 1984). The group 1 and 2 elements (Li, Na, K, Rb, Cs, Mg, Ca, Sr and Ba) tend to complex only with oxygen in ligands forming relatively unstable complexes and undergo rapid ligand exchange in water. These elements are generally very water soluble and only cations found in high concentration (Mg, Ca and Na) contribute significantly to the solid or sedimentary material. A variety of transition metals, including Mn II, Fe II, Co II, Ni II, Cu II, Zn II and Pb II demonstrate strong affinities for ligands containing O, N and S (Buffel, 1984).

Mn, Fe and Cu can easily be converted to other redox states with Fe III and Mn IV generally forming stable oxo or hydroxo complexes. Another group, containing Cu I, Tl I, Cd II, Ag I, Au I and Hg II, possess strong affinities for ligands containing N and S (Buffel, 1984). In natural systems, the concentration of these last two groups of transition metals is generally less than the total ligand concentration and therefore form stable insoluble complexes in water. The solubility of the various transition metals is related to the stability of soluble and insoluble ionic complexes formed with 0, N and S containing ligands.

Fresh water and sea water are dominated by different types of ligands. In fresh water, the ligand concentration varies immensely and often has higher concentrations of organic ligand than in marine waters. There are three major groups of organic ligands. First, freshly excreted products from primary and secondary producers including carbohydrates, proteins and lipids which are decomposed within days or weeks. Secondly, refractory organic matter which constitutes 70-80 % of the total dissolved organic matter. Refractory materials are primarily composed of aromatic hydrocarbons, chlorinated hydrocarbons and others which are stable to decomposition. Lastly, humic and fulvic acids which are polymorphous particulate complexes. Organic pollution can actually increase the ability of a fresh water system to adsorb excess metals and thereby lower the expression of effects by toxic metals. The concentration of ligands in sea water is relatively stable and dominated by inorganic ligands, primarily chloride, carbonate, borate, silicate, sulfate and ammonia (Jørgensen and Jensen, 1984). The concentration of soluble metal complexes is somewhat higher in fresh water than in salt water due to the greater abundance of organic ligands.

Ligand exchange and redox reactions can affect the solubility of chemical species in water, thereby influencing elemental mobility. As noted above, soluble species can be transported with bulk water movements. Recomplexation can result in either the solubilization of species from sediments or the precipitation of soluble species. Changes in physical parameters such as ph, pE, ionic strength or temperature can change the stability constants for the various complexes. In addition, variations in ligand concentrations can lead to competition for the metal ions, leading to recomplexation. In fresh water, the pH is generally about 7.0 and many metals form insoluble complexes with humic and fulvic acids, hydroxides and some carbonates. If the pH is lowered significantly, these insoluble complexes can reequilibrate to soluble or other insoluble forms. Al, Fe, Mn and Zn are all mobilized significantly from sediments under acidification induced by acid precipitation from fossil fuel combustion or drainage from mining operations. By contrast, as fresh water enters salt water, the inorganic ligand concentration increases dramatically. With the increases in chloride, sulfate and carbonate, many metal complexes exchange ligands and are either precipitated or solubilized in coastal regions. Natural waters are in dynamic flux due to normal water movements and inputs of compounds from terrestrial environments. These movements of water and fluxes of chemicals leads to reactions involving the precipitation and dissolution of other compounds already residing in the aquatic system.

The movement of elements depends upon the hydrodynamics of the waterway and the solubility of the various chemicals. Since the soluble metals diffuse through the water freely, these species can be transported with bulk movements of water. On the other hand, a large number of trace metals discussed above, form insoluble compounds. Most transition metals discharged into fresh water are bound to humic or fulvic acid sediments, whereas metal ions in marine systems generally precipitate as hydroxides, carbonates and sulfides or are bound to soils and sediments (Jørgensen and Jensen, 1984). Adsorption of metals onto sediments is generally fast, and can be modeled either by equilibrium terms or by first order kinetics. The adsorption of metals into clay is dominated by pH, with the clay acting as an ion exchange resin. Insoluble chemical species are relatively immobile but can be transported as particulates. The particulate and sediment content of rivers and streams has increased about three fold due to increases in mining, agriculture and industrial activity, with the distribution of metals in solution being approximately 17% dissolved and 72% as suspended particulates (Bacinni, 1984). The distance that sediments travel depends upon the size of the particle and the rate of flow in the waterway. Smaller particles travel greater distances before they settle to the bottom. These smaller particles also have a greater capacity for bound or ligated elements since they have a larger surface-to-weight ratio. Sediments are transported from the fast moving watercourses to lakes and estuaries where they reside and are buried by more siltation. The sediments at the bottom of lakes and estuarines become storehouses for these complexes.

Particulates and sediments can also undergo movement resulting from various physical disturbances. The influx of particles and movement of sediments in rivers often rises as a result of erosion, flooding or inputs of greater than normal amounts of water during storms or snow melts. These events can lead to physical changes including resuspension, settling and burial of particulates. In coastal regions, tidal movements also induce movement and redistribution of sedimentary material. Storm actions amplify wave activity which can dramatically increase erosion and movement of particulates. Once particles settle out of solution, there are random and cyclic events which result in mass movement and redistribution of materials.

The water in estuaries undergoes additional movement. The surface water, which is rich in organic nutrients and algal life, flows out of the basin towards thesea. The algae then settle to the bottom and the organic material is mineralized by other microorganisms. This water then flows back into the estuarial basin by a counter-current of water originating from the ocean depths. Often in Norwegian fjords, this deep current of water becomes anoxic due to respiratory utilization of oxygen (Stumm and Morgan, 1981). These estuarine currents and biological activities can affect the movement of not only dissolved elements but also those elements found in the sediments.

BIOLOGICAL INFLUENCE ON ELEMENTAL TRANSPORT

Organic and inorganic ligands have similarities and differences. Metal ions generally coordinate to unshared electron pairs on 0, S, N and Cl. Physical parameters in the aqueous environment will affect the stability of the metal ligand interaction with organic and inorganic ligands demonstrating similar trends. Due to the complexity and heterogeneity of many organic ligands, stability constants are harder to measure. In addition, complexation within living organisms can be controlled through the ability of cells to compartmentalize compounds and to control factors including ligand concentration, pH, pE and ionic strength within the cell. Cells can also contruct specific molecules designed for transport and storage of specific metals. Certain organisms can secrete chelator molecules which bind insoluble metals in the sediment (Wood and Wang, in press). These complexes can then specifically interact with the cell membrane and tansfert the metal to the cell where the metal can then ultimately be internalized and stored for use. By this or similar mechanisms, organisms can accumulate insoluble, essential elements thereby promoting elemental transport and mobility.

Metal ligand complexes undergo constant change. Metals bound to particulates or found in sediments undergo aging as crystal lattice structures reorganize into more stable forms. Both insoluble and soluble complexes undergo ligand exchange resulting in desorption/adsorption and dissolution/precipitation reactions. For example, proteinaceous ligands can be degraded, resulting in the release of the metal ion or a ligand can be protonated or deprotonated with concommitant shifts in stability and solubility of the metal ligand complex. Other changes in complexes result from ligand competition and changes in other physical parameters. Changes such as these can occur during the summer months as vast amounts of CO_2 are utilized by photosynthetic organisms. Typically, this activity leads to an increase pH from 7.2 to 9.5 which reflects a decrease in the CO_3^- concentration. Cd and Zn complexes are very sensitive to changes in pH in this region. Carbonate-metal complexes will also be affected by the utilization of CO_2. Alkalinity changes during photosynthesis because NO_3^-, NH_4^+ and HPO_4^{2-} are assimilated with conurrent uptake of H^+ or OH^- by the cells. Biological processes which affect alkalinity include photosynthesis, nitrate metabolism and sulfate metabolism (Table 3).

TABLE 3

Processes Affecting Alkalinity

Process	Alkalinity change for forward reaction
Photosynthetis and Respiration :	
(1a) $nCO_2 + nH_2O \underset{respir.}{\overset{photos.}{\rightleftharpoons}} (CH_2O)_n + nO_2$	NC
(1b) $106CO_2 + 16NO_3^- + HPO_4^{2-} + 122H_2O + 18H^+ \underset{respir.}{\overset{photos.}{\rightleftharpoons}} \{C_{106}H_{263}O_{110}N_{16}P_1\} + 138O_2$ "algae"	I
(1c) $106CO_2 + 16NH_4^+ + HPO_4^{2-} + 108H_2O \underset{respir.}{\overset{photos.}{\rightleftharpoons}} \{C_{106}H_{263}O_{110}N_{16}P_1\} + 107O_2 + 14H^+$	D
Nitrification :	
(2) $NH_4^+ + 2O_2$ ---------- $NO_3^- + H_2O + 2H^+$	D
Denitrification :	
(3) $5CH_2O + 4NO_3^- + 4H^+$ ---------- $5CO_2 + 2N_2 + 7H_2O$	I
Sulfide Oxidation :	
(4a) $HS^- + 2O_2$ ---------- $SO_4^{2-} + H^+$	D
(4b) $FeS_2(s) + 15/4O_2 + 3\ 1/2H_2O$ ---------- $Fe(OH)_3(s) + 4H^+ + 2SO^{2-}$ pyrite	D
Sulfate Reduction :	
(5) $SO_4^{2-} + 2CH_2O + H^+$ ---------- $2CO_2 + HS^- + H_2O$	I
$CaCO_3$ Dissolution :	
(6) $CaCO_3 + CO_2 + H_2O$ $Ca^{2+} + 2HCO_3^-$	I

Reprinted from Stumm and Morgan, 1981.
NC = no change ; D = decrease ; I = increase.

As mentioned earlier, pH changes occur from pollution stresses such as acid rain or mine runoff of from geochemical reactions such as dissolution of calcium carbonate. In acidified regions, species diversity and population density drop dramatically with some lakes and streams becoming devoid of life. Changes in metal complexes occur both naturally and biologically, with direct and indirect effects on ligand concentration and complex stability.

Besides metal ligand interactions and hydrodynamic movements of water, chemical equilibria need to be considered in understanding elemental movement through aquatic systems. Only a few elements, C, N, O, S, Fe and Mn, are predominant participants in aquatic redox processes (Stumm and Morgan, 1981). In an aquatic system at equilibrium with atmospheric oxygen, the reduction potential can be defined ($P(O_2)$ = 0.21 atm., E(H) = 800 mV at pH 7 and 25°C) (Stumm and Morgan, 1981). Using standard redox potentials (Table 4) for pertinent aquatic reactions, virtually all biochemically important elements should be found in their highest naturally occurring oxidation states : C as CO_2, HCO_3^- or CO_3^{2-}, N as NO_3^{2-}, S as SO_4^{2-}, Fe as FeOOH or Fe_2O_3 and Mn as MnO_2 (Stumm and Morgan, 1981). Even atmospheric nitrogen should be largely converted to nitrate under natural conditions according to thermodynamic constraints. Clearly, natural waters are far from thermodynamic equilibrium. Biological systems can extract energy for growth and division by catalyzing these thermodynamically favored and kinetically slow reactions.

Biological systems can be divided into two major classifications, photosynthetic and respiratory. Photosynthesis can be conceived as a process by which light energy is converted to chemical energy (Stumm and Morgan, 1981). The compounds formed are thermodynamically unstable and typically involve the conversion of compounds in a high oxidation state, C_2, to more reduced forms, CHO. This process generates a highly negative redox potential and a reservoir of O_2, a good oxidant. Respiration, on the other hand, tends to restore equilibrium by catalytic conversion of unstable compounds to thermodynamically favored forms in higher oxidation states. Energy is derived from this conversion which is used to drive reactions necessary for life. Organisms mediate these spontaneous reactions but cannot catalyse non-spontaneous reactions, though reaction conditions within the cell can be controlled so the thermodynamics become favorable at the expense of energy.

A sequence of redox reactions in an aqueous system can be generated to correspond to a thermodynamically feasible order. In a closed aquatic system containing organic material, oxidation occurs first by the reduction of oxygen then the reduction of nitrate, nitrite and manganese dioxide (if present), followed by FeOOH reduction to Fe(II) and at sufficiently negative redox potential, fermentation and reduction of carbon dioxide and sulfate may occur (Stumm and Morgan, 1981). These reactions follow a decrease in redox potential level, with the organism deriving more energy from aerobic respiration than from reactions using other electron acceptors. An ecological system follows this

TABLE 4

Equilibrium constants and redox processes pertinent in aquatic conditions (25°C).

Reaction	$p\varepsilon^o$(≡ log K)	$p\varepsilon^o(W)$ [a]
(1) 1/4 $O_2(g)$ + H^+ + e = 1/2 H_2O	+20.75	+13.75
(2) 1/5 NO_3 + 6/5H^+ + e = 1/10$N_2(g)$ + 3/5H_20	+21.05	+12.65
(3) 1/2 $MNO_2(s)$+1/2$HCO_3^-(10^{-3})$+3/2H^++e = 1/2$MnCO_3(s)$ +3/8H_2O	-	+ 3.90 [b]
(4) 1/2 NO_3^- + H^+ + e = 1/2NO_2^- + 1/2H_20	+14.15	+ 7.15
(5) 1/8 NO_3^- + 5/4H^+ + e = 1/8NH_4^+ + 3/8 H_20	+14.90	+ 6.15
(6) 1/6 NO_2^- + 4/3H^+ + e = 1/6 NH_4^+ + 1/3 H_20	+15.14	+ 5.82
(7) 1/2 CH_3OH + H^+ + e = 1/2$CH_4(g)$ + 1/2H_20	+ 9.88	+ 2.88
(8) 1/4 CH_20 + H^+ + e = 1/4$CH_4(g)$ + 1/4H_20	+ 6.94	- 0.06
(9) FeOOH(s) + $HCO_3^-(10^{-3})$ + 2H^+ + e = $FeCO_3$ + 2H_20	-	- 0.80 [b]
(10) 1/2 CH_20 + H^+ + E = 1/2 CH_3OH	+ 3.99	- 3.01
(11) 1/6 SO_4^- + 4/3H^+ + e = 1/6S(s) + 2/3H_20	+ 6.03	- 3.30
(12) 1/8 SO_4^{2-} + 5/4H^+ + e = 1/8$H_2S(g)$ + 1/2H_20	+ 5.25	- 3.50
(13) 1/8 SO_4^{2-} + 9/8H^+ + 3 = 1/8HS^- + 1/2H_20	+ 4.25	- 3.75
(14) 1/2 S(s) + H^+ + e = 1/2$H_2S(g)$	+ 2.89	- 4.11
(15) 1/8 $CO_2(g)$ + H^+ + e = 1/8$CH_4(g)$ + 1/4H_20	+ 2.87	- 4.13
(16) 1/6 $N_2(g)$ + 4/3H^+ + e = 1/3NH_4^+		
(17) 1/2 ($NADP^+$) + 1/2H^+ + e = 1/2 (NADPH)	- 2.0	- 5.50 [c]
(18) H^+ + e = 1/2 $H_2(g)$	O.O	- 7.00
(19) Oxidized ferredoxin + e = reduced ferredoxin	- 7.1	- 7.10 [d]
(20) 1/4 $CO_2(g)$ + H^+ + e = 1/24 (glucose) + 1/4H_20	- 0.20	- 7.20 [e]
(21) 1/2 $HCOO^-$ + 3/2H^+ + e = 1/2CH_20 = 1/2H_20	+ 2.82	- 7.68
(22) 1/4 $CO_2(g)$ + H^+ + e = 1/4CH_20 + 1/4H_20	- 1.20	- 8.20
(23) 1/2 $CO_2(g)$ + 1/2H^+ + e = 1/2$HCOO^-$	- 4.83	- 8.33

Reprinted from Stumm and Morgan, 1981

a Values for $p\varepsilon^o(W)$ apply to the electron activity for unit activities of oxidant and reductant in neutral water, that is, at pH = 7.0 for 25°C.

b These data correspond to (HCO_3^-) = 10^{-3} M rather than unity and so are not exactly $p\varepsilon^o(W)$; they represent typical aquatic conditions more nearly than $p\varepsilon^o(W)$ values do.

c M. Calvin and J.A. Bassham, The photosynthesis of carbon compounds, W.A. Benjamin, Menlo Park, Calif., 1962. NADP = Nicotinanide adenine dinucleotide phosphate.

d D.I. Arnon, Science 149, 1460 (1965)

e Lehninger, Bioenergetics, W.A. Benjamin, Menlo Park, Calif., 1965.

thermodynamic system fairly closely starting with aerobic heterotrophs and followed by denitrifiers, fermenters, sulfate reducers and finally methane bacteria. This sequence of biological mediated redox processes has been illustrated by Stumm (Fig. 3).

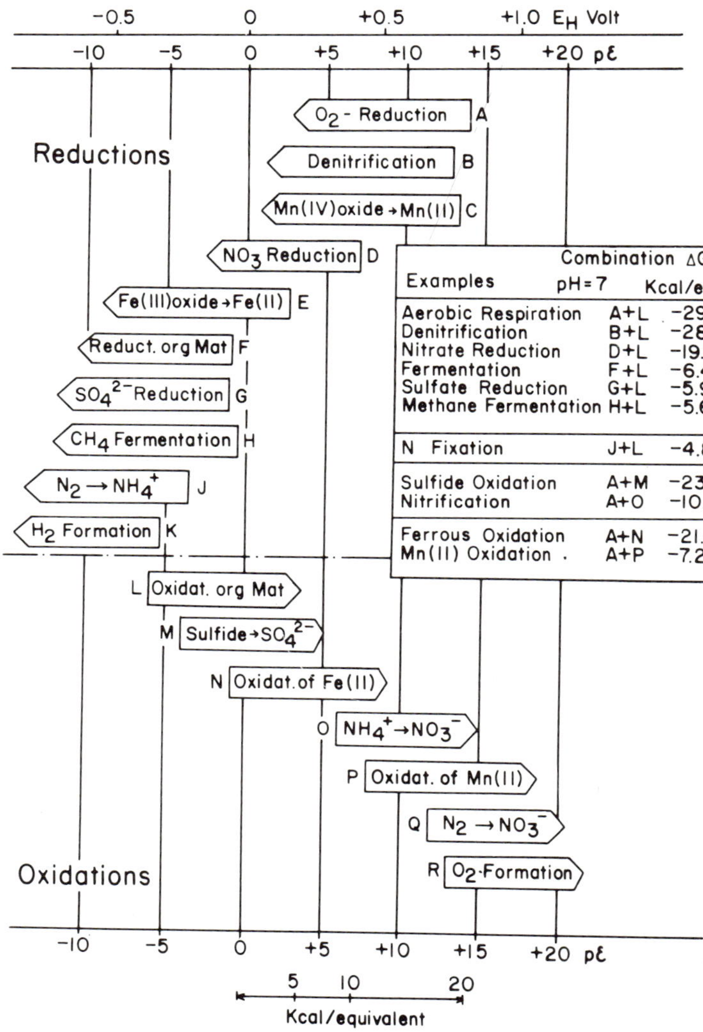

Fig. 3. Biologically mediated redox processes. Reprinted from Stumm and Morgan, 1981.

Microorganisms have adapted to efficiently harvest energy from kinetically slow but thermodynamically favorable reactions through the continuum of redox potentials found in nature.

Organisms will utilize compounds which provide the most energy under any set of conditions. This prnciple establishes the sequence of electron acceptors in redox processes to be O_2, NO_3^- and SO_4^{2-}. The ecological system above is generally found stratified in sediments. Metabolism requiring oxygen is found in the top 0.5 to 1.0 cm (Blackburn, 1983, Sorenson, 1984) of sediment as a consequence of permeability rates and the rate of utilization by microorganisms. The next zone is occupied predominantly by the nitrate reducers which is the next best electron acceptor. Once the nitrate is exhausted, and if the organic substrates are not themselves depleted, sulfate reduction predominates. The amount and complexity of organic substrates decrease with the depth of penetration into the sediments as a result of biological utilization. Anaerobic metabolism and depletion of the various substrates leads to a decrease in oxidation potential and the subsequent mobilization of metals from insoluble complexes. Regions of rivers or lakes high in nitrate, phosphate and organic material, can often become depleted of dissolved oxygen due to respiratory metabolism, resulting in anaerobic respiration and the subsequent mobilization of metal ions form the sediments. Mobilized metals will reprecipitate as insoluble complexes when the oxygen concentration increases. Typically, reoxygenation occurs within the sediments so the metals are reprecipitated without much movement, but when the water becomes anoxic, these metals can travel great distances before reoxygenation occurs. This mobilization-reprecipitation cycle can also lead to ligand exchange reactions. As a consequence, soluble complexes can also form is the appropriate ligands are available, such as chloride. In this way, mercury and cadmium can be solubilized from sediments under anaerobic conditions and recomplex to chlorine in marine estuaries.

Biological activity influences the mobilization and ligand exchange of metal complexes as a result of aerobic and anaerobic metabolism.Tidal action not only facilitates particulate movement but also results in rather dramatic changes in environmental conditions for the organisms living in those areas. Desulfovibrio is a sulfate reducing microroganism and is the dominant anaerobe in estuarial salt marshes, accounting for up to 80 % of the anaerobic population (Bartha, in press). When the salt marsh is flooded, this organism reduces sulfate to sulfide which results in the precipitation of mercury as mercuric sulfide. On the other hand, when the tide is out and the sulfate concentration is low, Desulfovibrio catalyzes the formation of methyl mercury under anaerobic conditions. Methyl mercury is highly permeable to cells and is mobilized into the biota. In this case, biological activity and elemental mobility are also influenced by cyclic changes, such as by tidal action. Similar changes in biological activity are seen for diurnal cycles, seasonal cycles and random effects resulting from weather.

Many of the reactions and interactions mentioned above occur in estuaries as fresh water mixes with salt water. The influx of fresh water brings sediments, particulates and dissolved nutrient materials. The salt water body provides increased concentrations on inorganic ligands which can compete with organic ligands bound to metals in the fresh water. Estuarial currents promote the breakdown of the various chemical, particulate and thermal gradients thereby affecting biological activity. Tidal movements result in periodic changes in the many chemical and physical gradients. Periodic flooding and storm activity dramatically increase material movement. Additional movements of sediments occur in many river basins as a result of dredging to keep water transportation routes navigable. Biotic and abiotic reactions can easily be dominated by the fresh water input of nutrients with estuaries often being regions of high biological activity. Since many large river estuaries are heavily populated by man, these systems are usually heavily used and polluted from sewage effluent, industrial activity and agricultural runoff.

REFERENCES

Baccini, P., 1984. Regulation of trace metal concentrations in freshwater systems. In: H. Sigel (ed.), Metal Ions in Biological Systems, Marcel Dekker, New York, Vol. 18, Ch. 8.

Bartha, 1985. Science, in press.

Blackburn, T.H., 1983. In: W.E. Krumben (ed.), Microbial Geochemistry. Blackwell Scientific Publications, Oxford, Ch. 3.

Buffel, J., 1984. Natural organic matter and metal-organic interactions in aquatic systems. In: H. Sigel (ed.), Metal Ions in Biological Systems, Marcel Dekker, New York, Vol. 18, Ch. 6.

Fortsner, U., Wittman, G.T.W., 1979. Metal pollution in the aquatic environment. Springer-Verlag, New York, Berlin.

Jørgenson, S.E. and Jensen, A., 1984. Process of metal ions in the environment. In: H. Sigel (ed.), Metal Ions in Biological Systems, Marcel Dekker, New York, Vol. 18, Ch. 3.

Lovelock, J.E., 1979. Gaea, A new look at life on earth. Oxford Univ. Press, Oxford, U.K.

Pearson, R., 1968. J. Chem. Educ., 45: 643.

Salomons, W., Baccini, P., 1984. The importance of chemical speciation in environmental processes. Dahlem Konferenzen, sept. 1-7. Berlin Springer-Verlag, in press.

Sorenson, J., 1984. Seasonal variation and control of oxygen, nitrate and sulfate respiration in coastal marine sediments. In: M.J. Klug and C.A. Reddy (eds), Current Perspectives in Microbial Ecology. American Society for Microbiology, Washington, D.C., pp. 447-453.

Stumm, W., Morgan, J.J., 1981. Aquatic Chemistry, 2nd Edition. John Wiley and Sons, New York.

Wood, J.M., 1984. Metabolic cycles for toxic elements in the environment. Science, 1983: 1049-1052.

Wood, J.M., 1984. Evolutionary aspects of metal ion transport. In: H. Sigel (ed.), Metal Ions in Biological Systems, Marcel Dekker, New York, Vol. 18, Ch. 7.

Wood, J.M. and Wang, H.K., 1983. Microbial resistance to heavy metals. Environ. Sci. Technol., 17(2): 582A-590A.

Wood, J.M. and Wang, H.K., 1985. Strategies for microbial resistance to heavy metals. In: W. Stumm (ed.), Chemical Processes in Lakes. Wiley Interscience, New York.